AF867733

Notes on Mathematics for Programming

Bobby Neal Winters

Cover images © Shutterstock.com

innovativeink
PUBLISHING

www.innovativeinkpublishing.com
Send all inquiries to:
4050 Westmark Drive
Dubuque, IA 52004-1840

Print ISBN: 979-8-3851-3165-5
Ebook ISBN: 979-8-3851-3166-2

Published in the United States of America

Contents

Chapter 1

Integers

1.1 Introduction

In this chapter, we will learn about the division algorithm and some of its applications. These applications will include: the modulus of a number and integer division; putting an integer N into a given base b; special things than can be done when the base b is a power of 2; modular arithmetic; divisibility tests including casting out nines, which you may know, and casting out sevens, which you may not; the method of encoding numbers into base 2 using ASCII code. We end the chapter by touching briefly on the mathematical principle of induction and recursive definition. While some explanation of why the mathematics works is essential, the focus of the chapter is on the teaching of calculation and the teaching of the language, and not on the proof.

1.2 The Division Algorithm

1.2.1 A Thought Experiment

Lets us begin with a thought experiment. Suppose that you are playing a game of cards with your friends. You have a deck of n cards and there are d players in the game. In this particular game, the practice is to deal out an equal number of cards to each player and to deal out cards until it is no longer possible to do so equally to each player.

Dealing cards is something most of us have done, but let's review the process. When you deal cards, in order to make sure everyone gets the same number of cards, you do it in rounds. On a given round, either you deal one card to each player or you don't deal to anyone. This requires the dealer to compare the number of un-dealt to the number of players before beginning a round.

As all of the cards are to be dealt out, the process will continue until you are unable to deal. This means you will come to a round where you cannot deal cards to every player. Let r be the number of cards that are left over. It is possible that all of the cards have been dealt out, in which case $r = 0$, but we know for certain that $r < d$ (recall d is the number of players) because if $r \geq d$, we would be able to deal another round.

When we are done, each player will have the same number of cards as every other player. Let's say this number is q. Given this, we can now express n, the original number of cards, in the following way:

$$n = dq + r$$

where $0 \leq r < d$.

See the video for this link here

1.2.2 The Division Algorithm

In (1.2.1), our thought experiment was a folksy version of what is called *the Division Algorithm*[1]. As a mathematician my first grown-up introduction to the Division Algorithm wasn't as an algorithm, but as a formula within a theorem:

Theorem 1.2.1. (The Division Algorithm) *Given a nonnegative integer n and a positive integer d there are unique nonnegative integers q and r with $0 \leq r < d$ such that*

$$n = d \cdot q + r. \tag{1.1}$$

As you see in this form, this is more a formula than an algorithm, but it does refer to the product of the algorithm–the integers q and r–and guarantees their uniqueness. Uniqueness is important because once it's guaranteed, one may find more convenient processes to obtain the numbers.

With notation as in Theorem 1.2.1, we will refer to n as the *numerator*; we will refer to d as the *denominator*. The number q is called the *quotient* and r is referred to as the *remainder*. Let us now take a few moments to show that quotient and remainder are unique for a given pair of numerator and denominator.

[1] We were, after all, *dividing* up cards among the players.

Suppose that

$$n = dq + r$$

and

$$n = dq' + r'$$

where q and r (and their primed versions) are as in Theorem 1.2.1. Then

$$dq + r = dq' + r'$$

so that

$$r - r' = dq' - dq = d(q' - q). \tag{1.2}$$

Since $0 \le r, r' < d$, it follows that $|r - r'| = 0$. This is because if $r - r' = d(q - q')$, then one of the following must be true: $|r - r'| = 0$, $|r - r'| = d$, $|r - r'| = 2d$, $|r - r'| = 3d$, *etc.* For only one of these is it true that $|r - r'| < d$, and that is the case $|r - r'| = 0$. Therefore, $r = r'$.

As we know that $r = r'$, Equation 1.2 becomes

$$d(q - q') = 0,$$

so either $d = 0$ or $q - q' = 0$. As d is positive, it follows that $q - q' = 0$ and as a consequence $q = q'$.

1.3 Modulus and Integer Division

Let us now that the mathematical results we obtained in Section 1.2.2 and put them into a form we will meet them in computer programming. We will express the division algorithm in terms of binary operations. Binary operations will be explored more thoroughly in Chapter 3.

In Section 1.2.2, we said that given a numerator and a denominator we can generate a quotient and a remainder. We can rephrase this in terms of a pair of binary operations: Integer divisions and modulus.

Let d be an integer with $d > 1$. Suppose that n is a nonnegative integer. The Division Algorithm says there are unique integers q and r with $0 \le r < d$ such that

$$n = dq + r.$$

We can express this by writing

$$q = n//d$$

and

$$r = n \,\%\, d.$$

The second of these is more often written as

$$r = n \mod d,$$

but we choose to express ourselves using the other notation because % and // are used sometimes in programming[2]. We call the operation // *integer division* and refer to q as the *quotient* (or *integer quotient* when we want to very precise). We say that r is n *modulo* d or $n \mod d$. We will find this way of characterizing the Division Algorithm to be useful.

1.4 A Number in a Base b

We will now begin one of the major tasks of this chapter as we learn how to render numbers into bases.

Consider the number 1729. We can express it as

$$1729 = 1 \cdot 10^3 + 7 \cdot 10^2 + 2 \cdot 10 + 9.$$

We can do it this way because our standard system for writing numbers is base 10. We use base 10 as an accident of history. Some suggest, we very good rationale, that base 12 would've been a better choice, but that ship has sailed. We are stuck with base 10 until such a time as we are defeated by 12-fingered alien overlords.

Any integer $b > 1$ can be used as a base and any positive integer can be expressed in any base, which we will show in this section. We will provide a well worked-out example in Section 1.4, as our exposition in this section is necessarily general. Let us begin by supposing that we have a number N expressed in a base b. Suppose

$$N = a_n b^n + a_{n-1} b^{n-1} + \cdots + a_1 b + a_0, \tag{1.3}$$

where $0 \le a_i < b$ for $i = 0, 1, \ldots, n$. We may shorten this to

$$N = (a_n a_{n-1} \ldots a_1 a_0)_b$$

[2]In particular, they are used in Python

or even

$$N = a_n a_{n-1} \dots a_1 a_0$$

in a context where b is understood. This is how we express N in base b.

In computer programming, we will primarily be interested in bases 2, 8, and 16 in addition to base 10 that we use in congress with our fellow human beings. We will want to be able to switch among these bases, so we will now construct a process by which we may render any number into any base.

Let us look back at Equation1.3. Using the Division Algorithm, we see

$$N = (a_n b^{n-1} + \cdots + a_1)b + a_0$$

so that

$$a_0 = N \,\%\, b$$

and

$$a_n b^{n-1} + \cdots + a_1 = N // b.$$

In order to prepare for an iterative process, define

$$N_0 = N // b$$

and

$$N_{k+1} = N_k // b$$

for $k \geq 0$. By following through the calculations, we can see that for $k \geq 0$,

$$a_{k+1} = N_k \,\%\, b.$$

What we just described is a method to put any positive number into any base $b > 1$. You could probably do these calculations by hand yourself at this point, but let me make it formal by putting it into a theorem.

Theorem 1.4.1. *Given a nonnegative integer N and an integer $b > 1$, let*

$$N_0 = N // b$$

and

$$a_0 = N \,\%\, b.$$

For each integer $k \geq 0$, let

$$N_{k+1} = N_k // b$$

and

$$a_{k+1} = N_k \,\%\, b.$$

Then there is an integer n such that $N_k = 0$ *for* $k > n$ *and so that*

$$N = a_n b^n + a_{n-1} b^{n-1} + \cdots + a_1 b + a_0.$$

While this is theorem, it is also an algorithm for expressing N in base b. I will now explain why it works.

We will first show that N_k must be equal to 0 for large values of k. First recall

$$N = bN_0 + a_0$$

so that

$$bN_0 = N - a_0.$$

As $a_0 \geq 0$, it must be that

$$bN_0 \leq N.$$

If we are in a situation where $N > 0$, since $b > 1$, it follows that

$$N_0 < N;$$

if $N = 0$, then $N_0 = N = 0$, but in this case putting N into any particular base isn't much of a chore. In the same way, we can show

$$N_{k+1} < N_k$$

whenever N_k is itself not equal to zero. If $N_k = 0$, all following values in the series will also equal zero.

Since we are dealing with integers, each reduction in equantity in the chain $N > N_0 > N_1 > \ldots$ is at least 1.[3] For $k \geq N$, we must have $N_k = 0$. Let n be the smallest value such that $N_n = 0$.

It is our claim that

$$N = a_n b^n + a_{n-1} b^{n-1} + \cdots + a_1 b + a_0.$$

We will show this for the case $n = 3$, *i.e.* $N_3 = 0$ but $N_2 \neq 0$. In this case

$$N = bN_0 + a_0 \quad (1.4)$$
$$N_0 = bN_1 + a_1 \quad (1.5)$$
$$N_1 = bN_2 + a_2 \quad (1.6)$$
$$N_2 = bN_3 + a_3 \quad (1.7)$$

[3]Indeed, we will be going down by a factor of b at each stage, but the extra precision would just be annoying in this context.

$//5$	$\% 5$
345	4
69	0
13	4
2	3
0	2

Table 1.1: Putting $N = 1729$ into base $b = 5$.

As $N_3 = 0$, we have $N_2 = a_3$. By substituting this into (1.6),

$$N_1 = b(a_3) + a_2 = a_3 b + a_2.$$

Taking this result and substituting it in turn into (1.5) yields

$$N_0 = b(a_3 b + a_2) + a_1 = a_3 b^2 + a_2 b + a_1.$$

Taking this and putting it into (1.4) will give us the desired result.

$$N = b(a_3 b^2 + a_2 b + a_1) + a_0 = a_3 b^3 + a_2 b^2 + a_1 b + a_0.$$

Example of Putting a Particular Number into a Particular Base

In order to better explain the results of Section 1.4, we will now work an example by putting $N = 1729$ into base $b = 5$. You can take a calculator and show

$$\frac{1729}{5} = 345.8,$$

so that $1729//5 = 345$. Now note that

$$1729 - 5 \cdot 345 = 4,$$

so that $1729 \% 5 = 4$. Now note that $\frac{345}{5} = 69$ exactly so that $345//5 = 69$ and $345 \% 5 = 0$. We can continue in this way until the process stops and put our results into Table 1.1. Displaying your calculations in a table like this is particularly handy as we can read off our number in base 5 from the bottom of the second column to the top. In this case

$$(1729)_{10} = (23404)_5.$$

We may check our result by calculating

$$4+5\cdot 5+4\cdot 5^2+3\cdot 5^3+2\cdot 5^4$$

is indeed equal to 1729.

For a worked out example on video check here

.

1.4.1 Integers Expressed in a Base $b = 2^n$

In this section, we will discuss one of the advantages of working in numbers expressed in a base $b = 2^n$. In order to do this, recall that given an integer $b > 1$ and a positive integer N, we may express N in the base b as in Equation 1.3, for some non negative integer n, and integers $0 \le a_i < b$ for $i = 0, \ldots, n$. We can abbreviate this by using a place-value notation writing $(a_n a_{n-1} \cdots a_1 a_0)_b$. Expressing a number in this way requires b distinct symbols as possible values for the a_i.

When we work in base 10, we use the set of symbols $\mathscr{S} = \{0,1,2,3,4,5,6,7,8,9\}$. If we were to work in base $b = 4$, we could use the subset $\{0,1,2,3\}$. For any base $b < 10$, we can use the obvious subset of $\mathscr{S}$.

For $b > 10$, the set $\mathscr{S}$ is no longer adequate for our needs as it simply doesn't have enough symbols. While we could come up with novel, alien numerals for our purposes, we have instead chosen to augment the set $\mathscr{S}$ with letters from the Roman alphabet. As we are going to be interested in base 16, the so-called *hexadecimal* form of numbers, we will use the set

$$\mathscr{H} = \{0,1,2,3,4,5,6,7,8,9,A,B,C,D,E,F\}$$

with the additional symbols corresponding to the values below

Symbol	Value (decimal)
A	10
B	11
C	12
D	13
E	14
F	15

In Section 1.4, we saw a method to put a number into a given base b. We will now see that is it very easy to change any number from base 2 to any base $b = 2^n$. Let us illustrate this by changing $N = (111101)_2$ into bases 4, 8, and 16.

Note that

$$\begin{aligned} N &= 1\cdot 2^5+1\cdot 2^4+1\cdot 2^3+1\cdot 2^2+0\cdot 2+1 \\ &= (1\cdot 2+1)2^4+(1\cdot 1)2^2+(0\cdot +1) \\ &= (1\cdot 2+1)4^2+(1\cdot 1)4+(0\cdot +1) \\ &= (11)_2 4^2+(11)_2 4+(01)_2. \end{aligned}$$

That is

$$N=(11)_2 4^2+(11)_2 4+(01)_2. \tag{1.8}$$

Now compare this to the binary form of N where the bits are grouped as pairs

$$N=(11\ 11\ 01)_2$$

and see the exact correspondence. We can now put the coefficients of Equation 1.8 into numerals from the set $\{0,1,2,3\}$ to see

$$\begin{aligned} N &= 3\cdot 4^2+3\cdot 4+1\cdot 1 \\ &= (331)_4. \end{aligned}$$

We group the bits in Equation 1.8 into pairs is because 4 (the base) is equal to 2^2, *i.e.* we use groups of 2 because the exponent is 2. To change N into base $8=2^3$, we will put the bits of N into groups of 3. This gives us

$$\begin{aligned} N &= (111\ 101)_2 \\ &= (111)_2\cdot 8+(101)_2\cdot 1 \\ &= 7\cdot 8+5\cdot 1 \\ &= (75)_8. \end{aligned}$$

In order to render N into base $16=2^4$, we will need to break the binary representation of N into groups of 4. Since the given binary representation of N only has 6 bits, we will pad its left end with two leading zeros.

$$\begin{aligned} N &= (00111101)_2 \\ &= (0011\ 1101)_2 \\ &= (0011)_2\cdot 16+(1101)_2\cdot 1 \\ &= 3\cdot 16+13\cdot 1 \\ &= (3D)_{16}. \end{aligned}$$

Note here that D is the hexadecimal symbol for $(13)_{10}$.

1.5 Modular Arithmetic

Having applied the Division Algorithm to render numbers into a given base b, we will now make an application of it in a different direction. We will now begin the process of defining and then applying modular arithmetic.

Modular Arithmetic is a way of doing arithmetic with numbers which have been evenly spaced in a cycle around a clock face, but we will not approach it that directly. We will first define a basic relation on the set of integers that we will call *divides* and then use this relation to define another relation on the integers called *equivalence modulo d* for a given d. This relation will give us numbers in a cycle around a clock face, as it were. Then we will learn how to do arithmetic on the "clock face numbers."

Let use now proceed to take the first of these steps.

1.5.1 Divides

Let a and d be integers. We say that d *divides* a if there is an integer n such that

$$a = d \cdot n.$$

In this case, we write $d|a$. On a practical level, when it comes to programming, one way we can test whether d divides a is by calculating $a \% d$. If $a \% d = 0$, then d divides a; otherwise it does not.

The divides relation has a number of useful properties that we will summarize in the following theorem.

Theorem 1.5.1.

1. *For every integer a, $a|a$.*

2. *For integers a, b, and c, if $a|b$ and $b|c$, then $a|c$.*

3. *For integers a, b, and c, if $a|b$ then $a|bc$.*

The divide relation works and plays well with both addition and multiplication. Suppose $d|a_1$ and $d|a_2$. Then $a_1 = dn_1$ and $a_2 = dn_2$ for some integers n_1 and n_2. So

$$\begin{aligned} a_1 + a_2 &= dn_1 + dn_2 \\ &= d(n_1 + n_2). \end{aligned}$$

Since $n_1 + n_2$ is an integer, $d|a_1 + a_2$.

Also note

$$a_1 a_2 = d n_1 d n_2 = d(d n_1 n_2).$$

Since $d n_1 n_2$ is an integer, $d|a_1 a2$. We summarize these two results in the following theorem.

Theorem 1.5.2. *Suppose a_1, a_2, and d are integers such that $d|a_1$ and $d|a_2$. Then*

1. $d|a_1 + a_2$ *and*
2. $d|a_1 a_2$.

1.5.2 Equivalence Modulo d

We will now define another relation on the integers called *equivalence modulo d*, where $d > 1$ us an integer. Equivalence modulo d is an example of an equivalence relation[4]. Equivalence relations are very useful and you will learn about them in great generality should your educational pursuits take you in that direction.

Let us now concentrate on this specific equivalence relation. Let $d > 1$ be an integer. Suppose that n_1 and n_2 are integers. We say that n_1 *is equivalent to n_2 modulo d* if $d|(n_1 - n_2)$. Though this is usually denoted

$$n_1 = n_2 \mod d,$$

we will denote it as

$$n_1 \equiv_d n_2.$$

Theorem 1.5.3. *Let $d > 1$ be an integer.*

1. *For every integer n, $n \equiv_d n$.*
2. *For integers n_1 and n_2, if $n_1 \equiv_d n_2$, then $n_2 \equiv_d n_1$.*
3. *For integers n_1, n_2, and n_3, if $n_1 \equiv_d n_2$ and $n_2 \equiv_d n_3$, then $n_1 \equiv_d n_3$.*

[4]It's right there in the name.

The three properties in Theorem 1.5.3 are the very ones that make $\equiv_d$ an equivalence relation, by the way.

To see that (1) is true, note $d|(n-n)$ because $n-n=0=d\cdot 0$, and 0 is an integer.

Let us now consider (2). Suppose $n_1 \equiv_d n_2$. Then $d|(n_1-n_2)$. Therefore, $d|(-1)(n_1-n_2)$, *i.e.* $d|(n_2-n_1)$. This is equivalent to saying $n_2 \equiv_d n_1$.

Finally, we look at (3). Note that $d|(n_1-n_2)$ and $d|(n_2-n_3)$ because $n_1 \equiv_d n_2$ and $n_2 \equiv_d n_3$. Consider

$$n_1 - n_3 = (n_1-n_2)+(n_2-n_3).$$

Since $d|(n_1-n_2)+(n_2+n_3)$ it follows $d|(n_1-n_3)$, that is to say $n_1 \equiv_d n_3$.

There is one more very important thing we need to make note of. Suppose $r = n \,\%\, d$ for integers n and d with $d>1$. Then

$$n = d\cdot(n//d)+r.$$

From this we see

$$n-r = d\cdot(n//d).$$

This is to say that $d|n-r$. As a consequence, $n \equiv_d r$. This will allow us to simplify our arithmetic greatly when we get into the thick of modular arithmetic. We make this formal in the following theorem.

Theorem 1.5.4. *Let n and d be integers with $d>1$. Let $r = n \,\%\, d$. Then $n \equiv_d r$.*

After all this, you might want to check out this video

.

1.5.3 An Example

We are in danger of drifting off into a La-La Land of formulas, so let's bring some numbers onto the stage. Consider Table 1.2.

Note that the columns $n \,\%\, 3$ and $n \,\%\, 5$ cycle through the numbers $0,1,2$ and $0,1,2,3,4$, respectively. There are numbers we could wrap cyclically around clockfaces, as it were.

Let's now do a couple of calculations in order to see how $\equiv_d$ acts around the arithmetic operations.

n	$n \% 3$	$n \% 5$
0	0	0
1	1	1
2	2	2
3	0	3
4	1	4
5	2	0
6	0	1
7	1	2
8	2	3
9	0	4
10	1	0

Table 1.2: Some integers mod 3 and 5

Let's add 4 to itself in a couple of different ways and see if we get anything that makes sense. First note $4 \equiv_3 1$ and that $1+1=2$. Now note that $4+4=8$ and $8 \equiv_3 2$. We get the same answer each time. That's promising.

Let's try this again with different numbers: 2 and 5. Note that on one hand $2+5=7 \equiv_3 1$. On the other hand $2 \equiv_3 2$ and $5 \equiv_3 2$. Then $2+2=4 \equiv_3 1$. Again we get the same answer each time.

This is evidence that $\equiv_d$ respects addition.

Let's now try with multiplication. Let's try 2 times 3. Note that $2 \cdot 3 = 6 \equiv_3 = 0$. Now $2 \equiv_3 2$ of course and $3 \equiv_3 0$, and $2 \cdot 0 = 0$, so $\equiv_3$ seems to respect multiplication.

In (1.5.4), we will show this is not an accident.

1.5.4 Modular Arithmetic

In this section, we will show that equivalence modulo d respects addition and multiplication. We state this precisely in the following theorem.

Theorem 1.5.5. *Let $d > 1$ be an integer and let a, a', b, and b' be integers such that $a \equiv_d a'$ and $b \equiv_d b'$. Then*

1. $a + b \equiv_d a' + b'$ *and*

2. $ab \equiv_d a'b'$.

It is not difficult to see that these claims are true. Consider

$$(a+b)-(a'+b')=(a-a')+(b-b').$$

Since d divides both terms of the righthand side, it will divide the righthand side, and therefore, of course, the lefthand side. Therefore, $a+b \equiv_d a'+b'$.

To see the other claim, only requires a bit more stealth. Note

$$\begin{aligned} ab-a'b' &= ab-ab'+ab'-a'b' \\ &= a(b-b')+b'(a-a'). \end{aligned}$$

Note that d divides each term of the second line of the righthand side, so it must be that d divides $ab-a'b'$. It follows that $ab \equiv_d$.

For some more worked-out examples look at this video

.

1.6 Applications

The material we've been covering may all be very interesting and is very basic to the portion of mathematics we call number theory, but we are learning it because it is useful. We will dig in to a lengthy application in Chapter 2. In the meantime, we will look at some quicker applications.

1.6.1 Casting Out Nines

Consider the number $N = 19,479,673,294$. Here's a question: Can we tell whether N is evenly divisible by 9 without actually dividing by 9?

There is a method taught in school–called "Casting Out Nines"–which answers this question. The idea is to take the *digit sum* of N and use the fact that the digit sum of N is divisible by 9 exactly when N is divisible by 9. The digit sum is well-named because it is just the sum of the digits. For our particular N, the digit sum is 61. I know enough arithmetic to see that 61 is not divisible by 9, but if I didn't I could use the fact that $6+1=7$ and that 7 is not divisible by 9. If you don't know that 7 is not divisible by 9 I don't now how to help you.

This method works because we've rendered N into base 10 and because of modular arithmetic. In particular, it works because $10 \equiv_9 1$ so that

$$100 = 10 \cdot 10 \equiv_9 1 \cdot 1 \equiv_9 1,$$

and indeed $10^n \equiv_9 1$ for $n = 1, 2, \ldots$. Therefore,

$$a_n 10^n + \cdots + a_1 10 + a_0 \equiv_9 a_n + \cdots + a_1 + a_0,$$

which is the digit sum. Indeed, as $9 \equiv_9 0$, it is not even necessary to include any a_i that is equal to 9 in our sum. We can also exclude groups of digits whose sum is nine. In practice, we just look at these and mark them out. Hence the name "Casting out Nines." Sometimes the names make sense.

A bonus to this technique is that–in addition to knowing whether the given number is divisible by 9, we can know what the remainder is when divided by 9. Given the N of this section, $N \% 9 = 7$.

The see this done, look at this video

.

1.6.2 Casting Out Sevens

In Section 1.6.1, we learned "Casting out nines," a technique for checking divisibility by 9. There are also techniques for checking divisibility by 2, 3, 4, 5, 6, and 8. In looking at this list of digits, one might note the conspicuous absence of 7. The truth is that one can use modular arithmetic to develop such a test, but if we are honest, it is almost as simple to just divide by 7 and get it over with. That is to say, the test for divisibility is awkward *if we are given numbers in base 10.*

If we are lucky enough to have a number in base 8, testing for divisibility 7 is as easy as testing for divisibility by 9 when we have a number in base 10. It is the same process as casting out nines, changing what obviously needs to be changed. We can call it casting out sevens if we like.

Consider the number $N = (37461)_8$. Then

$$N = 3 \cdot 8^4 + 7 \cdot 8^3 + 4 \cdot 8^2 + 6 \cdot 8 + 1.$$

Now $8 \equiv_7 = 1$, $8^2 \equiv_7 1$, and $8^n \equiv_7 1$ for integers $n \geq 1$. So

$$\begin{aligned} N &\equiv_7 3+7+4+6+1 \\ &\equiv_7 3+4+6+1 \\ &\equiv_7 0. \end{aligned}$$

So N is divisible by 7.

As you might imagine, this is not very practical for ordinary hand-written work because changing from base 10 to base 8 is more difficult that dividing by 7. The situation is different if we have our numbers in base 2 as we would in a computer.

As we saw in Section 1.4.1, we can change from base 2 to base 8 quite easily. Consider

$$\begin{aligned} M &= (11011000001)_2 \\ &= (011\ 011\ 000\ 001)_2 \\ &= (3301)_8 \\ &\equiv_7 0. \end{aligned}$$

So M is divisible by 7.

To see this done, look at this video

.

1.6.3 Encoding Numbers in Binary

This section veers less into mathematics and more into computers than we have done so far. It impinges upon one of the most difficult aspects of computer programming: Input/Output.

When we program a computer, we have to translate from human language to computer language. At the most basic level, this involves translating "human numbers" into "computer numbers." It is commonly known that computers store numbers in binary form for the most part. This raises the question: How do we convert from decimal to binary? The attentive reader must be forgiven for thinking we've already done that for we have not. We've hidden a layer of complexity.

The user does not enter numbers into the computer; he enters *text*. There are often numerals embedded with in this text, but they are text none the less. The text itself breaks down into *characters*, and each character is encoded as a number. So the numbers aren't numbers, they are text. The text is composed of characters, and each character is a number, but *not* the number is was to begin with. Who could possibly be confused?

One method of encoding characters is the American Standard Code for Information Interchange. This is denoted by the acronym ASCII which is pronounced as "ass-kee." The ASCII code establishes a correspondence between a set of characters and control sequences on on hand and the numbers 0 through

	0000	0001	0010	0011	0100	0101	0110	0111
0000	[NA]	[NA]	[NA]	0	[NA]	P	[NA]	p
0001	[NA]	[NA]	[NA]	1	A	Q	a	q
0010	[NA]	[NA]	[NA]	2	B	R	b	r
0011	[NA]	[NA]	[NA]	3	C	S	c	s
0100	[NA]	[NA]	[NA]	4	D	T	d	t
0101	[NA]	[NA]	[NA]	5	E	U	e	u
0110	[NA]	[NA]	[NA]	6	F	V	f	v
0111	[NA]	[NA]	[NA]	7	G	W	g	w
1000	[NA]	[NA]	[NA]	8	H	X	h	x
1001	[NA]	[NA]	[NA]	9	I	Y	i	y
1010	[NA]	[NA]	[NA]	[NA]	J	Z	j	z
1011	[NA]	[NA]	[NA]	[NA]	K	[NA]	k	[NA]
1100	[NA]	[NA]	[NA]	[NA]	L	[NA]	l	[NA]
1101	[NA]	[NA]	[NA]	[NA]	M	[NA]	m	[NA]
1110	[NA]	[NA]	[NA]	[NA]	N	[NA]	n	[NA]
1111	[NA]	[NA]	[NA]	[NA]	O	[NA]	o	[NA]

Table 1.3: Partial ASCII Table

127 on the other. In the Table 1.3, we present only the digits, uppercase letters, and lowercase letters; these are the so-called *alphanumeric* characters. We've left out the special characters and the codes that were intended to control the flow of text; most of the latter are obsolete at this code was created during the era of teletype[5]. We denote the omitted characters by [NA] as they are not applicable to our current concern.

The Table 1.3 is quite cleverly designed. The rows and columns are labeled by the binary numbers $(r_3r_2r_1r_0)_2$ and $(c_3c_2c_1c_0)_2$, respectively. The table is laid-out so that the character in column $(c_3c_2c_1c_0)_2$ and row $(r_3r_2r_1r_0)_2$ has the ASCII code $(c_3c_2c_1c_0\ r_3r_2r_1r_0)_2$.

For example, the character 0 (zero) has the ASCII code $(0011\ 0000)_2$. Note

[5]Look it up

that

$$\begin{aligned}(0011\ 0000)_2 &= (0011)_2 \cdot 16 + (0000)_2 \cdot 1 \\ &= 3 \cdot 16 + 0 \cdot 1 \\ &= 48\end{aligned}$$

Similarly, the code for the character 9 (nine) is

$$\begin{aligned}(0011\ 1001)_2 &= (0011)_2 \cdot 16 + (1001)_2 \cdot 1 \\ &= 3 \cdot 16 + 9 \cdot 1 \\ &= 57\end{aligned}$$

A careful examination of Table 1.3 will reveal that the ASCII code for the character for the digit d is

$$N = 48 + d. \tag{1.9}$$

We abuse notation here by allowing d to be the digit as well as the value of the digit.

From this we can see that entering the digit 3 on your keyboard will result in it being encoded as

$$48 + 3 = 51 = (0011\ 0011)_2.$$

This formula can be rewritten as

$$d = N - 48,$$

which is to say that if we know a give number N is the ASCII code for a digit, we can recover that digit by subtracting 48 from it.

Very rarely will we be dealing with numbers that are single digits. Most numbers will be entered as a string of text. For example, suppose that the computer is going to use the number $N = 1729$. How does it change the text 1729 into it binary equivalent?

First each digit will be converted from its ASCII code to its value as above. Once we do that, recall that the number itself is equal to

$$1729 = 1 \cdot 10^3 + 7 \cdot 10^2 + 2 \cdot 10 + 9.$$

The idea is that there will be a routine set up in the computer that performs this calculation. This requires the computer to perform nine multiplications, and multiplication is relatively time consuming when compared to addition, for instance. We can rearrange this so that fewer multiplications are required. Consider

$$\begin{aligned}
1729 &= 1\cdot 10^3+7\cdot 10^2+2\cdot 10+9\\
&= (1\cdot 10^2+7\cdot 10+2)\cdot 10+9\\
&= ((1\cdot 10+7)\cdot 10+2)\cdot 10+9\\
&= (((0001)_2\cdot(1010)_2+(0111)_2)\cdot(1010)_2+(0010)_2)\cdot(1010)_2+(1001)_2
\end{aligned}$$

Note that the final form only requires the computer to perform 3 multiplications.

The observant reader who was really tuned-in during basic algebra might see that we've performed these operations much as one evaluates a polynomial in so-called synthetic division. Mathematicians refer to this method as *Horner's Method* or *Horner's Algorithm*. Indeed, consider the polynomial

$$p(x) = a_n x^n + a_{n-1}x^{n-1} + \cdots + a_1 x + a_0.$$

If $b > 1$ is an integer and $0 \leq a_i < b$ for $n = 0, 1, \ldots, n$, it follows that$p(b) = (a_n \ldots a_1 a_0)_b$.

If you need more help, look at this video.

1.7 The Principle of Induction and Recursive Definition

This section is about the Principle of Induction and the closely related concept of recursive definition. Before we venture into this abstract realm, let us begin with a good, solid metaphor.

Consider an array of dominoes. By this I mean an arrangement of dominoes wherein if one domino is knocked over, the rest will fall in turn. These dominoes have to be arranged very carefully. There is one domino that is distinguished from the rest by virtue of being first: When it is flicked over, the rest will fall in turn. When one is setting up a domino array, one must be mindful it is done in such a way so that when any given domino falls, the one that is next will fall too.

This is very similar to the Principle of Induction. It is used to prove propositions $P(n)$ where n is an element of set of consecutive integers that is bounded below. For sake of simplicity of exposition, we will state the definitions in terms of the set $\{0, 1, 2, 3, \ldots\}$ but the method works for any set of consecutive integers that has a smallest element whether that element is 0, a positive number, or a negative number.

Having set out a suitable metaphor, let's create the mathematical notation so that we may be precise.

For an integer $n \geq 0$, let $P(n)$ be a mathematical proposition. Suppose

1. $P(0)$ is true and
2. for every $k \geq 0$ such that $P(k)$ is true it follows that $P(k+1)$ is true,

then $P(n)$ is true for every $n \geq 0$. In terms of our domino metaphor (2) is saying the dominoes are placed in such a way so that one knocks down the next and (1) is saying the first domino is knocked down. As compelling as this metaphor is, it is not a proof of the Principle of Induction. The Principle of Induction can be proven using the *Least Integer Principle*. This states that every set of integers that is bounded below has a least element. The proof is not hard, but it would lead us too far afield.

The Principle of Induction is often used to prove statements about functions or concepts that are defined recursively. Many computer programs make use of recursion. At this point, it is easier to give examples of recursion than to state a rigorous definition.

For a non negative number n, we can define n *factorial*–written as $n!$–by saying $n!$ is the product of n and all of the positive integers less than n, *e.g.*

$$\begin{aligned} 3! &= 3\cdot 2\cdot 1 \\ &= 6. \end{aligned}$$

This method is quite intuitive and works well until we consider $0!$–which we will have to–then it makes no sense. We can remedy this by defining *factorial* recursively in the following way.

First define

$$0! = 1,$$

then, for $n \geq 0$ define

$$(n+1)! = (n+1)\cdot n!$$

By this new definition

$$\begin{aligned} 3! &= 3\cdot 2! \\ &= 3\cdot(2\cdot 1!) \\ &= 3\cdot(2\cdot(1\cdot 0!)) \\ &= 6. \end{aligned}$$

We may define other well-know mathematical expressions recursively as well. For example, given a real number $x \neq 0$, define

$$x^0 = 1$$

and, for $n \geq 0$, define

$$x^{n+1} = x \cdot x^n.$$

Given the two preceding recursive definitions, we can prove the follow result by using the Principle of Induction.

Proposition 1.7.1. *For $n \geq 4$,*

$$2^n < n!$$

Let's now use the Principle of Induction to prove this. We can by hand, using the recursive definitions, calculate that $2^4 = 16$ and $4! = 24$. So the proposition is true for $n = 4$.

Now suppose that $k \geq 4$ is an integer such that

$$2^k < k!$$

We claim that

$$2^{k+1} < (k+1)!$$

Note

$$\begin{aligned} 2^{k+1} &= 2 \cdot 2^k \\ &< 2 \cdot (k+1) \cdot k! \\ &= (k+1)! \end{aligned}$$

Therefore, by the Principle of Induction,

$$2^n < n!$$

for $n \geq 4$.

1.8 Exercises for Chapter 1

1. As in Section 1.2.2, let n denote the numerator; d, the denominator; q, the quotient; and r, the remainder. In each of the following, you will be given n and d. Please find q and r.

   (a) $n = 12, d = 5$

   (b) $n = 15, d = 2$

   (c) $n = 25, d = 2$

   (d) $n = 25, d = 7$

   (e) $n = 154, d = 2$

   (f) $n = 153, d = 9$

2. As in Section 1.2.2, let n denote the numerator; d, the denominator; q, the quotient; and r, the remainder. In each of the following, you will be given q and r. Please find all possible pairs (n, d) such that $0 \le n \le 50$. Your answer may be empty.

   (a) $q = 12, r = 15$

   (b) $q = 24, r = 15$

   (c) $q = 10, r = 3$

   (d) $q = 7, r = 25$

3. As in Section 1.2.2, let n denote the numerator; d, the denominator; q, the quotient; and r, the remainder. In each of the following, you will be given n and r. Please find all possible pairs (d, q).

   (a) $n = 12, r = 2$

   (b) $n = 15, r = 2$

   (c) $n = 17, r = 5$

   (d) $n = 99, r = 9$

   (e) $n = 108, r = 0$

4. Perform the following calculations.

   (a) $15//5$

(b) 15//3

(c) 15//2

(d) 15//10

(e) 15//7

5. Perform the following calculations.

   (a) 15 % 5

   (b) 15 % 3

   (c) 15 % 2

   (d) 15 % 10

   (e) 15 % 7

6. Perform the following calculations.

   (a) 10 % 2

   (b) 15 % 2

   (c) 13 % 2

   (d) 15432 % 2

   (e) 15213117 % 2

7. Perform the following calculations.

   (a) 100 % 10

   (b) 15 % 10

   (c) 13 % 10

   (d) 15432 % 10

   (e) 15213117 % 100

8. Present the number $N = 625$ in base $b = 3$.

9. Present the number $N = 65$ in base $b = 5$.

10. Present the number $N = 77$ in base $b = 17$.

11. Given the number $N = 625$, present it in each of the following bases.

(a) $b = 16$

(b) $b = 8$

(c) $b = 2$

12. Given the number $N = 65$, present it in each of the following bases.

    (a) $b = 16$

    (b) $b = 8$

    (c) $b = 2$

13. Given the number $N = 73$, present it in each of the following bases.

    (a) $b = 16$

    (b) $b = 8$

    (c) $b = 2$

14. Given the binary number $(11001110001)_2$ express it in

    (a) $b = 8$

    (b) $b = 16$

15. Given the binary number $(10001110011)_2$ express it in

    (a) $b = 8$

    (b) $b = 16$

16. Given the binary number $(11001110001)_2$ express it in

    (a) $b = 8$

    (b) $b = 16$

17. The hexadecimal number $FF00FF$ is kind of pink in the RGB palette. Express it in

    (a) $b = 2$

    (b) $b = 8$

    (c) $b = 10$

18. The hexadecimal number $(0041FF)_{16}$ is kind of blue in the RGB palette. Express it in

    (a) $b = 2$

    (b) $b = 8$

    (c) $b = 10$

19. For each of the following numbers a, give all of the numbers d such that $d|a$.

    (a) $a = 12$

    (b) $a = 51$

    (c) $a = 120$

    (d) $a = 641$

20. Suppose $a = 337225065214$ and $b = 1475071467824$. It is a fact that 17 divides a and 17 divides b. Without calculating $a+b$ or performing division, use results from the book to justify that 17 divides $a + b$.

21. With n and d as given below, find r such that $0 \le r < d$ and $n \equiv_d r$.

    (a) $n = 123$, $d = 10$

    (b) $n = 123$, $d = 12$

    (c) $n = 51$, $d = 17$

    (d) $n = 51$, $d = 15$

22. For each of the following perform the indicated addition and give the answer modulo d. That is $(a + b) \equiv_d r$ where $0 \le r < d$.

    (a) $(11 + 4) \equiv_{12} r$

    (b) $(10 + 10) \equiv_{12} r$

    (c) $(5 + 4) \equiv_7 r$

    (d) $(4 + 6) \equiv_7 r$

23. For each of the following numbers, determine whether or not they are divisible by 9.

    (a) 198210981

(b) 45111330000

(c) 45111330007

(d) 70101033154

24. Calculate each of the following

(a) 198210981 % 9

(b) 45111330000 % 9

(c) 45111330007 % 9

(d) 70101033154 % 9

25. Given the following numbers in base 8, determine whether they are divisible by 7.

(a) $(776134)_8$

(b) $(436177)_8$

(c) $(11111111)_8$

(d) $(17000001)_8$

26. Calculate the ASCII codes for the following digits. Give both the binary and decimal form of the answers.

(a) 4

(b) 5

(c) 7

27. Calculate the ASCII codes for the following characters. Give both the binary and decimal form of the answers.

(a) H

(b) E

(c) L

(d) P

28. Use Horner's Method to evaluate the following polynomials at the given points.

(a) $x = 3$, $p(x) = 3x^4 + 2x^3 - 12x^2 + x + 7$

(b) $x = -2$, $p(x) = x^5 - 2x^4 - 2x^3 - 12x^2 + x + 7$

(c) $x = -1$, $p(x) = 3x^5 + 2x^3 - 12x^2 + x + 7$

29. Show $\dfrac{1}{n!} < \dfrac{1}{2^n}$ for $n \geq 4$.

30. Use the Principle of Induction to show $2^n < 3^n$ for $n \geq 1$.

Chapter 2

Calendar Time

2.1 Days of the Week

2.1.1 Some Illustrative Examples

Let us begin by creating a table that creates a correspondence between the days of the week and the integers from 0 to 6, inclusive. (See Table 2.1) The choice of the range of integers in this table has to do with modular arithmetic as was introduced in Section 1.5.4.

We will attempt to be very careful in our use of words. We will, in particular, be careful to make the distinction between "day" and "date." By itself, "day" means a particular 24-hour interval of time running from midnight to midnight. We can use it in the context of "day of the week" where it refers to one from a list of seven names: Sunday, Monday, Tuesday, Wednesday, Thursday, Friday, and Saturday. We might also refer to a "day of the month" where it would be referring to an appropriate ordinal number in the interval 1 to n, where $n = 28, 30$, or 31.

When we use the word "date," we are referring to a particular day within a fixed system of temporal nomenclature. For example, December 7, 1941 is the date associated to the day of a particular event in history. Note that the date

Sunday	Monday	Tuesday	Wednesday	Thursday	Friday	Saturday
0	1	2	3	4	5	6

Table 2.1: Days of the Week

referred to above consists of three parts: month, day, and year. When the year is understood, we will refer to the same date as December 7. During the month of December 1941, would have referred to it as "the seventh."

To begin our discussion of finding the which day of the week a particular date falls on, note that 7 days after any given date is the same day of the week, *e.g.* 7 days after Tuesday is Tuesday again. We can represent this mathematically in reference to Table 2.1 as

$$x + 7 \equiv_7 x.$$

However, we can do more that this. For example, if today is Tuesday (which corresponds to 2) 9 days from now will be

$$\begin{aligned} 2+9 &\equiv_7 11 \\ &\equiv_7 4 \end{aligned}$$

which corresponds to Thursday on our table. Likewise, 17 days from this Tuesday is

$$\begin{aligned} 2+17 &= 19 \\ &\equiv_7 5 \end{aligned}$$

which is Friday's number.

Let's make this a bit more concrete. July 4, 2021 was on Sunday. We will now figure out the days of the week some other days fell on that year without necessarily referring to a calendar.

We will start out easy. What day of the week did July 31 fall on? Note 31 - 4 = 27. Now

$$\begin{aligned} 0+27 &= 27 && (2.1) \\ &\equiv_7 6, && (2.2) \end{aligned}$$

so July 31 fell on Saturday.

That was too easy. To make it harder, let's ask the same question about August 16. This is harder because we've gone to a different month, but let's pretend that July got hungry and ate August.[1] More formally, let's say that we concatenated the two months and retained the name July for the composite month. So

[1]Math joke for children: Why is 6 afraid of 7? Because 7 8 9!

August 16 could be thought of as July 47 ($31 + 16 = 47$). There are then $47 - 4 = 43$ days that separate July 4 and August 16. As $43 \equiv_7 1$, it follows that August 16 is on Monday.

Let us now become more bold. What day of the week did Christmas fall on in 2021? This requires concatenation July (31 days), August (31), September (30), October (31), and November (30) to the top of December. Then December 25 is the 178th ($31 + 31 + 30 + 31 + 30 + 25$) day of July. So 174 ($= 178 - 4$) days separate the 4th of July and Christmas. As

$$174 \equiv_7 6,$$

it follows December 25, 2021 is on Saturday.

2.1.2 Outline of the General Method

In Section 2.1.1, while we worked out the day of the week of each date separately, we were actually applying a general method:

1. Find a base date with known day of the week that occurs before the date you are interested in. (To simplify, find a base date that is a Sunday and falls on January 1.)

2. Calculate the number of days between that base date and the date of interest.

3. Add that sum to the number associated with the day of the week of the base day. (This is why it's simpler if the day is Sunday.)

4. Take that sum modulo 7.

Reducing numbers modulo 7 is not particularly difficult. As we saw in Section 2.1.1, the most difficult part of the procedure is part (2) determining the number of days between the date of interest and the base date in a general way. We will now proceed to spend some substantial space to working out how to accomplish part (2) in a systematic way.

2.2 Days of the Year

We've already seen an *ad hoc* method for counting the days between a pair of dates. In our pursuit of a more general method, we will create a function

Month Number	Month Name	Standard Year Length	Standard Year Cumulative (σ)	Leap Year Length	Leap Year Cumulative (σ)
1	January	31	0	31	0
2	February	28	31	29	31
3	March	31	59	31	60
4	April	30	90	30	91
5	May	31	120	31	121
6	June	30	151	30	152
7	July	31	181	31	182
8	August	31	212	31	213
9	September	30	243	30	244
10	October	31	273	31	274
11	November	30	304	30	305
12	December	31	334	31	335

Table 2.2: Lengths and Cumulative in Standard/Leap Years

ord(date). The function **ord** takes a date as an input and returns its ordinal position within the year as an output, *e.g.* **ord**(January 5) = 5 because January 5 is the fifth day of the year and **ord**(February 2) = 33 because February 2 is the 33rd day of the year.

This calculation is not difficult in principle. The ordinal position of a date within its year is the sum of the day of the month with the total sum of the lengths of all the preceding months. In order to make this more precise, let us establish some notation.

Let the ordered triple (d, m, y) denote the dth day of the mth month of the year y. So $m/d/y$ becomes (d, m, y). For example, April 5, 1925 is denoted as $(5, 4, 1925)$. We will refer to months by their numbers when it is convenient to do so. Consider Table 2.2. An examination of this table reveals the length of a month is actually a function of two variables–the month and the year–with February changing from 28 to 29 days, depending upon whether or not the year is a standard year or leap year, respectively.

Let $\lambda(m, y)$ denote the length of month m in year y. Given a date (d, m, y), its ordinal position within the year y can be calculated as

$$\mathbf{ord}(d, m, y) = d + \sum_{k=1}^{m-1} \lambda(k, y). \tag{2.3}$$

We have included columns of cumulative values of month length in Table 2.2. We will put

$$\sigma(m, y) = \sum_{k=1}^{m-1} \lambda(k, y).$$

For example, when y is a standard year, $\sigma(5, y) = 120$; when y is a leap year, $\sigma(5, y) = 121$. Use of this formula will allow us to simplify Equation (2.3) to become

$$\mathbf{ord}(d, m, y) = d + \sigma(m, y). \qquad (2.4)$$

This is a simple formula, but the presence of y, the year, raises the question of whether the given year y is a leap year. We will address this presently.

You can see the video at this link.

2.3 Leap Years and the Determination Thereof

Let us begin this section with a deep dive on the word "year."

The word "year" is ambiguous. It could refer to a Solar (or Tropical) Year which is the number of days from one vernal equinox to the next; or it could mean the number of days required for the Earth do go completely around a circle in a coordinate system established with the sphere of fixed stars, *i.e.* the Sidereal Year.

Because of the importance of keeping track of the seasons, we chose to synchronize our calendar with respect to the Solar Year. This synchronization is complicated by the fact that the Solar Year (or the Sidereal Year for that matter) does not consist of an exactly integral number of days. To five decimal places, the Solar Year is 365.24217 days long. There are probably more decimal places, but these five will give us enough trouble.

What goes wrong if we ignore this fraction? Nothing immediately, but suppose that this year spring began on March 21. In 4 years, if nothing is done and we keep to a 365 day year, spring will start on March 22; in forty years it would start on March 31; in four hundred years spring would start 97 days later. This is a whole season off.

In order to remedy this, during the time of Julius Caesar, it was decided to add a day to the calendar every 4 years. These years are called *leap years*; we can call the added days *leap days*. The effect of this was to make the average length of calendar years to be 365.25 days. This is only a deviation of 0.00783

```
is_leap(year):
    if not (4|year) then:
        return False
    else if not (100|year) then:
        return True
    else if not (400|year) then:
        return False
    else:
        return True
```

Figure 2.1: Leap Year Algorithm

from our five-decimal value. This difference looks small, but if we use this value for 100 years or season sill be a day off; for 1000 years, 10 days off.

For this reason, we do another correction. Every 100 years, we will fail to add the leap day. This will give us an average year length of 365.24. This is 0.00217 from our five-decimal value, and this isn't really very much of an improvement. Because of this, we will add even another correction. Though every 100 years we fail to add a leap day, every 400 years we make decide to add it anyway. This will make the average length of the calendar year be 365.2425 days. This is 0.00033 from our five-decimal value. In 1000 years, this will be eight hours off; in 10,000 years, it will be three days off. This is what managers like to call "someone else's problem."

2.4 Counting Leap Years

In Section 2.3, we implicitly described an algorithm to determine whether a year is a leap year. We will now describe it explicitly in terms of the pseudo-code in Figure 2.1. Pseudo-code is a way of writing plans for portions of computer programs. The idea is to use language that is terse but whose meaning is discernible. Programs in pseudo-code are meant to be easily translatable into real computing languages.

In this section, we will learn how to count the number of leap years that are within a particular span of years. In order to do this, we will first consider a more abstract problem. In this context, we use the notation $[n, m)$ to be the set of integers (as opposed to real numbers) which are greater than or equal to n

but less than m.

Let n and M be positive integers. How many integers are in the interval $[1, M)$ (*i.e.* 1 is included but M is not) are divisible by n? For example, suppose $n = 3$ and $M = 9$. The integers in $[1,9)$ are $1,2,3,4,5,6,7,8$. Those divisible by 3 are 3 and 6, so there are 2 of them. A moment's reflection will reveal that this is equal to $8//3$. This leads us to the following.

Proposition 2.4.1. *Given positive integers n and M, the number of integers in the interval* $[1, M)$ *divisible by n is equal to* $(M-1)//n$.

Having obtained this general result, we can now apply it to obtain a formula for the number of leap years that occur *before* the year y:

$$\textbf{leap\_years}(y) = (y-1)//4 - (y-1)//100 + (y-1)//400. \tag{2.5}$$

Warning: It is very important to note that $m//n \neq m \cdot (1//n)$ for most values of m.

By this formula, the number of leap years at or after the year 1AD and before 1804AD is

$$\begin{aligned} \textbf{leap\_years}(1804) &= 1803//4 - 1803//100 + 1803//400 \\ &= 436 \end{aligned}$$

theoretical leap years. We say theoretical because this calendar, the Gregorian Calendar, was not introduced until 1582AD.

A video is available here.

2.5 Days of Our Years

In Section 2.1.1, we used *ad hoc* methods to determine the day of the week for some given dates within a single calendar year. In Section 2.1.2, we outlined a general method for doing this which involved setting a base date and calculating the number of days from that base date to the date of interest. We promised we would have more details for that general method, and now we will provide them.

Our first step will be to set (d_B, m_B, y_B) as the base date for our calendar. By *calendar*, I mean a list of days embedded in time; one day comes before the next. Not a hard concept. We use the notation (d, m, y) that we established earlier to denote these dates for the days in the calendar. Years, months, and days are simply a way we label dates, a way for mere humans to keep track of them. We will be interested in associating each of our dates (d, m, y) with a unique natural number $\textbf{Ord}(d, m, y)$. This will be done in such a way as to preserve the order of the dates, *i.e.*

1. $\textbf{Ord}(d_B, m_B, y_B) = 1$ and
2. if (d', m', y') is the day that immediately follows (d, m, y), then

$$\textbf{Ord}(d', m', y') = \textbf{Ord}(d, m, y) + 1.$$

With the mathematics we've developed up to this point, we are on the verge of being able to write down a formula for $\textbf{Ord}(d, m.y)$. First recall that in Section 2.2 we defined $\textbf{ord}(d, m, y)$ (note the lowercase "o") to be the order the date (d, m, y) falls within the year y. Note in particular Equation 2.4.

Now let $v(y)$ be the total number of days in the years $y_B, y_B+1, \ldots, y-2, y-1$, the years from our base year to the year just before the year y. As the number of days from (d_B, m_B, y_B) to (d, m, y) is equal to the days in the numbers of years in the calendar before the year y plus the number of days up to the date in the year y, we see

$$\textbf{Ord}(d, m, y) = \textbf{ord}(d, m, y) + v(y). \tag{2.6}$$

As we can calculate $\textbf{ord}(d, m, y)$ from Equation 2.4 using Table 2.2, we now need only a way to calculate $v(y)$ in order to find $\textbf{Ord}(d, m, y)$. Let L be the

number of leap years that occur in the interval $[y_B, y)$. Then L is equal to the number of leap days within those years. Therefore,

$$v(y) = 365 \cdot (y - y_B) + L.$$

We can calculate L by using Equation 2.5 in the following way:

$$\begin{aligned} L &= \textbf{leap\_years}(y) - \textbf{leap\_years}(y_B) \\ &= (y-1)//4 - (y-1)//100 + (y-1)//400 \\ &- (y_B-1)//4 + (y_B-1)//100 - (y_B-1)//400. \end{aligned}$$

Therefore,

$$\begin{aligned} \textbf{Ord}(d, m, y) &= \textbf{ord}(d, m, y) + 365 \cdot (y - y_B) \\ &+ (y-1)//4 - (y-1)//100 + (y-1)//400 \\ &- (y_B-1)//4 + (y_B-1)//100 - (y_B-1)//400. \end{aligned}$$

Recall now that $\mathbf{Ord}(d_B, m_B, y_B) = 1$, we can now calculate that the number of days between (d_B, m_B, y_B) and (d, m, y) is

$$\mathbf{Ord}(d, m, y) - \mathbf{Ord}(d_B, m_B, y_B) = \mathbf{Ord}(d, m, y) - 1. \tag{2.7}$$

A video that summarizes this can be found here.

2.5.1 An Example

For our example, we will set $(1, 1, 1804)$ as our base date. This date was chosen for two reasons. First, it falls after the date in which the Gregorian Calendar was adopted; second, it falls on Sunday whose day number is zero. We now choose a date to test our general method: $(7, 12, 1941)$.

To calculate $\mathbf{Ord}(7, 12, 1941)$, we will need to calculate both $\mathbf{ord}(7, 12, 1941)$ and $\nu(1941)$. First let us note that 1941 is a standard year. Now recall

$$\mathbf{ord}(7, 12, 1941) = 7 + \sigma(12, 1941).$$

From Table 2.2, we see that $\sigma(12, 1941) = 334$. From this it follows $\mathbf{ord}(7, 12, 1941) = 341$.

The calculation of $\nu(1941)$ is more involved, but still not difficult. First note $y - y_B = 1941 - 1804 = 137$. From this, it follows $365 \cdot (y - y_B) = 50005$.

Let us now calculate

$$\begin{aligned} \mathbf{leap\_years}(1941) &= 1940//4 - 1940//100 + 1940//400 \\ &= 470 \end{aligned}$$

In Section 2.4, we calculated $\mathbf{leap\_years}(1804) = 436$. We can now calculate

$$\begin{aligned} \nu(1941) &= 50005 + (470 - 436) \\ &= 50039. \end{aligned}$$

Therefore,

$$\begin{aligned} \mathbf{Ord}(7, 12, 1941) &= 341 + 50039 \\ &= 50380. \end{aligned}$$

Referring to Equation 2.7, we see

$$\mathbf{Ord}(7, 12, 1941) - \mathbf{Ord}(1, 1, 1804) = 50379.$$

These calculations are summarized in Table

As $50379 \ \% \ 7 = 0$, it follows that $(7, 12, 1941)$ fell on a Sunday. This can be checked by looking up Pearl Harbor in an American History book.

A video example is to be found here.

Regular Days	=365(1941 - 1804)		=	50005
	1941	1804		
$(y-1)//4$	485	450		
$(y-1)//100$	-19	-18		
$(y-1)//400$	+4	+4		
Leap Days	471	436		+34
$\sigma(12, 1941)$	$= 334$			
ord$(7, 12, 1941)$	$= 7 + \sigma(12, 1941)$		=	+341
				50380

Table 2.3: Calculations

2.6 Another Way of Looking at It

If you've read the previous sections of this chapter, a question may have occurred to you: Would it be possible to have a calendar unmarked by months and years, consisting only of an ordered list of days? That is to say that we would refer to each date only by a number. This would be reminiscent of the "Star Date" used in the *Star Trek* science fiction franchise. The answer is, of course, we can, and that it is a nice system for computers to use.

One issue, perhaps the main issue, associated with doing this is that it is not what humans are used to, and humans pay the electric bill. When the robots ascend, the system of ordinal days might be forced upon us, whether we like it or not, but in the meantime, we will need to translate between the usual system and the ordinal system.

We have spent the bulk of this chapter translating from dates to ordinal days. Let us now spend some time in the opposite direction.

2.6.1 What Year Is It?

For the sake of discussion, let us suppose we are using an ordinal calendar that is based on January 1, 1804 as we discussion in Section 2.5.1. In that calendar, the date January 1, 1804 corresponds to the ordinal number 1 and–because 1804 was a leap year–December 31, 1804 corresponds to 366. Note that January 1, 1805 corresponds to $367 = 1 + 366$ and December 31, 1805 corresponds to $731 = 366 + 365$ because 1805 is not a leap year.

It is not difficult to imagine that, as the years pass, the ordinal number as-

```
My_Year(ordinal_day)
    while ordinal_day > length(current_year):
        current_year = current_year + 1
        ordinal_day = ordinal_day − length(current_year)
    return current_year, ordinal_day
```

Figure 2.2: The **My_Year** Function

sociated with a particular day would become quite large. Suppose we find ourselves in a spaceship and wake up one day to find the ordinal day is 100,000. What date would it be in our old-style human calendar?

While the method of finding the answer to this is not as "mathematical" as we found going in the other direction, it is not difficult. We may proceed as in the function defined in Figure 2.2. This function refers to the function **length**(y) which returns the number of days of the year y depending upon whether or not it is a leap year. Note that we have one input into the function ordinal_day and two outputs: current_year and ordinal_day. The ordinal_day is used in two different ways. The version that is input is the ordinal day within the set calendar; the value of ordinal_day that is output has been corrected so that it is the ordinal day within the current_year that is returned by the function as well.

2.6.2 What Month is It?

In Section 2.6.1, Figure 2.2 gave us a function that took an ordinal day with a calendar and returned the year that ordinal day belonged in as well as the day's ordinal position within that year. We now wish to take that information and determine the day's date within the year. We now wish to take that information and determine the day's date within the year. Consider the function described in Figure 2.3. This function uses the function **length**(m, y) which returns the length of the month m in the year y. This function will return the month that contains the ordinal day and the day of the month.

2.7 What time is it and Why?

The purpose of this section is to calculate the time in part of the world when the time is known in another part of the world. Let us begin at the beginning: A

```
My_Month(ordinal_day, current_year)
    current_month = 1
    while ordinal_day > length(current_month, current_year)
        ordinal_day = ordinal_day − length(current_month, current_year)
        current_month = current_month + 1
    return current_month, ordinal_day
```

Figure 2.3: The **My_Month** Function

day has twenty-four hours.

The reason for the 24-hour day goes back to the Egyptians and Babylonians. Like us they divided the day into the light part and the dark part. Each the day and the night was separated into 12 hours. One might well ask, "Why twelve?" The author does not know. However, the following is interesting: It is natural to think of the path the sun cuts through the sky as being half of a circle. This semicircle has 180° in it. It is comparatively easy to use compass and straight edge, the height of technology in those days, to divide that semicircle into 12 equal arcs[2].

Over the course of the millennia, before the advent of any comprehensive, worldwide timekeeping system, the convention developed that 12 noon was the time of day when the sun was at its highest point in the sky, *i.e.* the sun is at its *zenith*. This is important because it is no doubt the reason that when such a comprehensive system developed, it was done in such a way as to preserve the sun being near its zenith at 12 noon.

Indeed, we could have one time for the whole world, but the price for this would be that when the clock read 12 noon in London with the sun at its zenith there, it would read 12 noon in New York and it would still be dark with sunrise still an hour away[3].

The fact is we do have one time for the whole world. It is called, *coordinated universal time.* Because the countries involved in setting this up, spoke different languages, a compromise was reached at it is referred to UTC. This is not an acronym in either of the languages. Neither got its way.

[2]We leave this as an exercise to enrich your inner math-nerd.

[3]At certain times of year, give or take.

2.7.1 Coordinated Universal Time

We resolved the problem of keeping a coherent time system and keeping the sun near its zenith at 12 noon around the world by creating time zones that are roughly parallel to lines of longitude. Mathematically, the time zones should be exactly parallel to lines of longitude, but this is one of those situations where politic trumps even mathematics and astronomy.

Each time zone is given a name of the form

$$\text{UTC} \pm h,$$

where the h (the UTC *offset*) is the number of hours the given time zone is either ahead $(+h)$ or behind $(-h)$ the base timezone, which for historical reasons is located around the Royal Observatory at Greenwich, England[4].

In Table 2.4, we list the timezone of a region when it is at standard time; we list a major city or identifying geographical locations within the zone; the letter associated with the zone in the alternative nautical system. This table is not complete but will be sufficient from which to learn calculations.

2.7.2 Calculating Time Differences

The first and easiest problem we will deal with is the calculation of time difference between two timezones. In our treatment, we will allow h to assume negative values so that we can name all zones in the form $\text{UTC} + h$ instead of $\text{UTC} \pm h$.

Given time zones $\text{UTC} + h_1$ and $\text{UTC} + h_2$. The time difference is $|h_2 - h_1|$. If $h_2 - h_1$ is positive, we say $\text{UTC} + h_2$ is $|h_2 - h_1|$ hours ahead of $\text{UTC} + h_1$. Otherwise, we say $\text{UTC} + h_2$ is $|h_2 - h_1|$ hours behind $\text{UTC} + h_1$

2.7.3 Simple Examples

What is the time difference between London (UTC+0) and New York (UTC−5)? We calculate that $0 - (-5) = 5$ so, London is 5 hours ahead of New York.

What is the time difference between Mexico City (UTC−6) and Cairo (UTC+2)? We calculate that $-6 - 2 = -8$, so Mexico City is 8 hours behind Cairo.

---

[4]If you find this unsatisfactory, I suggest you go back in time 300 years and solve all of the navigational problems that Great Britain did when it ruled the seas.

Standard Times

UTC offset	Example	Nautical
-12	Baker Island	Y
-11	American Samoa	X
-10	Honolulu	W
-9	Anchorage	V
-8	Los Angeles	U
-7	Denver	T
-6	Mexico City	S
-5	New York	R
-4	Santiago	Q
-3	Sao Paulo	P
-2	small islands	O
-1	Azores	N
0	London	Z
1	Berlin	A
2	Cairo	B
3	Moscow	C
4	Dubai	D
5	Karachi	E
6	Bangladesh	F
7	Jakarta	G
8	Shanghai	H
9	Tokyo	I
10	Sydney	K
11	small islands	L
12	Auckland	M

Table 2.4: Time Zones at Standard Time

2.7.4 The International Dateline

As one crosses the Pacific Ocean, one eventually passes over a geographical-political line called *the international dateline.* Travelers going from east to west will go forward one day when they cross, but will lose one hour; those going from west to east will lose a day while gaining an hour[5].

Let us now consider a traveler from Mexico City ($UTC - 6$) who is traveling to Tokyo ($UTC + 9$). Tokyo is 15 hours ahead of Mexico City. Suppose it is 6pm on Saturday in Tokyo when our traveler's plane lands there. What time does his body think it is?

For these calculations, it saves us much confusion if we work in a military-style 24-hour clock. In this system, 6pm is 18.00 hours. Since Tokyo is 15 hours ahead of Mexico City, Mexico City is 15 hours behind. So if is 3.00 hours on Saturday, that is to say, 3am on Saturday.

Suppose now that the traveler eats dinner, showers (trying his best to wash off the filth of a long, nonstop flight), sleeps, and eats breakfast. It is now 9:30am on Sunday. What is the time and day of week in Mexico City. Note that the time $9:30 = 9 + \frac{30}{60} = 9.50$. As Mexico City is 15 hours behind Tokyo, the time in Mexico City is $9.50 - 15 = -5.5$. We are not familiar with negative times on a clock, no one is. We can remedy this by taking -5.5 modulo 24. This can be done by adding 24 to the negative number, *i.e.* $-5.5 \% 24 - 5.5 + 24 \% 2418.5$. This can be read as $18:30$ hours or 6:30pm. Since the value we calculated was negative (-5.5), that will be on the previous day, which is to say, Saturday.

2.7.5 Seasonal Variations and the Equator

To complicate things further, at certain times of the year in order to compensate for changes in the length of daylight hours, whole regions elect to "spring forward" to daylight savings time or "fall back" to standard time. The act of springing forward is the mathematical equivalent to adding an hour to your UTC offset, *i.e.* $UTC + h$ shifts to $UTC + (h+1)$. Again we are in a situation where allowing h to be negative works with us.

Let us consider the time difference between Chicago ($UTC - 6$) and São Paulo ($UTC - 3$). One might hope this would be easy, but they are on opposite sides of the equator, and therefore, most of the time, when one is on daylight savings time, the other is not. During July, for example, Chicago is on daylight

[5]This puts the mathematician in me in mind of Riemann Surfaces, but that way madness lies

savings time, but São Paulo is not. That means Chicago is at $UTC + (-6 + 1) = UTC - 5$ during July so that Sã0 Paulo is two hours ahead of Chicago.

However, during January this is reversed. São Paulo is at daylight savings time which is $UTC + (-3 + 1)UTC - 2$ so that São Paulo is 4 hours ahead of Chicago.

2.8 Exercises for Chapter 2

1. If today is the day of the week given, what is the day of the week after the indicated number of days has passed?
   (a) If today is Sunday, 2 days from now will be what?
   (b) If today is Sunday, 12 days from now will be what?
   (c) If today is Saturday, 2 days from now will be what?
   (d) If today is Saturday, 12 days from now will be what?
   (e) If today is Wednesday, 102 days from now will be what?
2. September 11, 2001 fell on a Tuesday. On what days of the week did the following days fall that year?
   (a) Halloween, *i.e.* October 31.
   (b) Christmas, *i.e.* December 25.
   (c) New Year's Eve, *i.e.* December 31.
   (d) Saint Patrick's Day of 2002, *i.e.* March 17, 2002
   (e) July 4, 2002.
3. Why is 6 afraid of 7?
4. For a given year, answer each of the following.
   (a) How many days is July 2 after July 1?
   (b) How many days is August 7 after July 1?
   (c) How many days is October 31 after July 1?
   (d) How many days is it from March 21 to September 21?
   (e) Why don't we need to know whether or not it is a leap year in the above?
5. Answer each of the following.
   (a) In a leap year, how many days are there from September 21 to March 21?
   (b) In a non-leap year, how many days are there from September 21 to March 21?

6. During a year when Easter falls on April 4, what day of the week does Christmas fall on?

7. During a regular year, calculate **ord** for each of the following dates.

   (a) March 17.

   (b) June 16.

   (c) July 4.

   (d) September 11.

   (e) November 11.

8. During a regular year, calculate how many days it is from January 7 to each of the following dates.

   (a) March 17.

   (b) June 16.

   (c) July 4.

   (d) September 11.

   (e) November 11.

9. During a regular year, how many days is it from June 16 to November 11?

10. Determine which of the following years are leap years.

    (a) 1901

    (b) 1942

    (c) 1962

    (d) 1972

11. Which of the following are leap years?

    (a) 1900

    (b) 2000

    (c) 2200

    (d) 2400

12. List all of the leap years from 1950 to 2001, inclusive.

13. Find the number of multiples of 7 in each of the following intervals.

    (a) $[1,15)$
    (b) $[1,99)$
    (c) $[1,27)$
    (d) $[15,27)$
    (e) $[27,99)$

14. In the following intervals, find the number of multiples of 3 that are not multiples of 5.

    (a) $[1,109)$
    (b) $[1,49)$
    (c) $[1,87)$
    (d) $[87,109)$
    (e) $[49,87)$

15. How many theoretical leap years are there in the following intervals?

    (a) $[1,1500)$
    (b) $[1,1800)$
    (c) $[1,1573)$
    (d) $[1500,1573)$
    (e) $[1573,1800)$

16. Set the base date as January 1, 1972.

    (a) How many Gregorian leap years are there from the year 1 until, but not including, the base year?
    (b) How many Gregorial leap lears are there from the base year until, but no including, 2072?

17. Set the base data at January 1, 1804. Calculate the following.

    (a) $\mathbf{Ord}(15,4,1865)$
    (b) $\mathbf{Ord}(7,1,1941)$

(c) **Ord**$(22, 11, 1963)$

(d) **Ord**$(4, 4, 1968)$

(e) **Ord**$(6, 1, 2021)$

18. What is the time difference between the timezone UTC − 6 and each of the following timezones? Phrase your answers in terms of hours ahead and hours behind.

    (a) UTC − 8.

    (b) UTC − 12.

    (c) UTC + 6.

    (d) UTC + 4.

19. A traveler gets off a plane in Moscow UTC + 3 at 9pm on Monday. He needs to make phone calls to the cities below. Find the time and day of the week for each of the given cities.

    (a) Denver UTC − 7.

    (b) Tokyo UTC + 9.

    (c) Dallas UTC − 6.

    (d) Honolulu UTC − 10

20. When is midnight on New Year's Eve in Asuncion, Paraguay (UTC − 4), what time is it in Kansas City, Missouri (UTC − 6).

21. (**Extra Credit**) Write a computer program that uses the results of §§ 2.6.1 and 2.6.2. To calculate the current date from the ordinal day given a particular base date.

Chapter 3

Sets, Logic, and Counting

3.1 Sets

3.1.1 Introduction

Sets are a basic concept of mathematics. Sets are, in fact, essential to the usual approach taken in setting up the foundations of mathematics. It would be easy–and perhaps a great deal of fun for the author–to run down a rabbit hole here and explore sets in fullness. We will studiously avoid this.

A student who is able to read this text likely already has a solid, intuitive idea of what a set is and what its properties are. For the current purpose, this intuitive understanding will probably be adequate. Nonetheless, we will endeavor to be careful.

Before we go much further, let's look at some examples. Consider the two sets A and B as given: $A = \{1,2,3,4,5\}$ and $B = \{2,3,4,\}$ We say that 1,2,3,4, and 5 are the elements of A. We write $1 \in A$, $2 \in A$, *etc.* Not that B has elements 2,3, and 4. As these are all elements of A as well, we will say that B is a *subset* of A. In the coming section we will talk about this more carefully. It is expected that not of this is new to the reader. We have an intuition of sets and the language of sets is already well established in school mathematics.

3.1.2 Basic Concepts

As we now begin to define times, we will first posit a *univerisal set* U. The set U is *not* the set of everything. [1] Its purpose is to set a context which limits the sets we will be working with. The set U could be the set of all employees of a company; it could be the set of all stem bolts at a particular deep space station. Most of the time we will be talking about sets of numbers of a particular time, and knowing U will help us to exclude extraneous information.

If x is an element of U, we will write $x \in U$. We say S is a *subset* of U if every element of S is also an element of U. Given subsets S and T of U, we say T is a subset of S if every element of T is also an element of S. In this case we write, $T \subset S$.

There is a way of construction sets that is vary common. One can speak of the set of elements that has a particular property P. The notation to describe this is called *set-builder notation*, and appears as below

$$S = \{x \in U | x \text{ has property } P\}.$$

We read this as "the set S of elements x of U such that x has property P.

There is a unique subset of U that contains no elements. It is called the *empty set* and is denoted by $\emptyset$ or by $\{\}$. The empty set is a subset of every subset of U.

You might consider looking at the video here.

3.1.3 Examples of Sets

The Natural Numbers

The *natural numbers* denoted by $\mathbf{N}$ are the most basic set of numbers we will be working with. From an abstract point of view, we are interested in two properties of the natural numbers:

1. There is a $1 \in \mathbf{N}$ such that $1 \leq n$ for every $n \in \mathbf{N}$.

2. For every $n \in \mathbf{N}$, $n + 1 \in \mathbf{N}$.

---

[1] The "set of everything" is a big problem. Russell's Paradox, which is well beyond the scope of these notes, lets R be the set of all sets that aren't elements of themselves. Is $R \in R$? If so, then it's not. If not, then it is. Rather than think about that, let's make sure the question doesn't come up.

It is typical to denote the natural numbers as $\mathbf{N} = \{1,2,3,\ldots\}$ where the three dots indicate the numbers continue without end. The set **N** is the paradigm example of an infinite set.

Finite Sets

Having given **N** as an example of an infinite set, let us now look at a family of finite sets. For every $n \in \mathbf{N}$, let $\mathbf{N}_n$ be the set of elements k of **N** such that $k \le n$. In this case we write

$$\mathbf{N}_n = \{1,2,3,\ldots,n\}$$

where the three dots indicate the missing elements between 3 and n. So $\mathbf{N}_3 = \{1,2,3\}$ and $\mathbf{N}_5 = \{1,2,3,4,5\}$.

One can easily observe that if $n \le m$, then $N_n = N_m$.

It almost goes without saying–but we will say it anyway–that $\mathbf{N}_n$ contains exactly n elements.

Boolean Numbers

We will now introduce the reader to a different set of numbers, the *Boolean numbers* 0 and 1. We will denote the set of Boolean numbers as $\mathbf{Z}_2 = \{0,1\}$.[2]

We recognize 0 and 1 as being bits, binary digits, from our expansion of numbers base 2. In electrical engineering applications, 0 can be used to denote "off" and 1 to denote "on." We will soon be using 0 to denote logical **False** and 1 to denote logical **True**.

We will let $\mathbf{Z}_2$ to be ordered by $<$ with $0 < 1$ as one would expect.

3.2 Functions

3.2.1 Introduction

The reader has no doubt be introduced to the concept of functions in an earlier course. At the introductory level, a function is typically defined by a formula such as

$$f(x) = x^2 + 1,$$

where x represents a number. At this level, on often speaks of the *domain* of the function as being the set of numbers one is allowed to "plug in" for x and still

[2] The reason for this notation can be found in a course on abstract algebra.

have the formula make sense and the *range* as being the set of possible values that $f(x)$ might assume. We must now move beyond these naive notions in order to prepare our minds for the needs of computer programming.

3.2.2 The Abstract Notion of Function

The use of functions in computer programming requires a more abstract notion than simple formulas can provide. We will view a function as a "black box." This is meant to convey a mechanism where something goes in one end and comes out the other. The *domain* can be any set and does not necessarily have to contain any numbers at all. We will abandon the term range entirely as being linguistically problematic in favor of the term *co-domain.* The purpose of the co-domain is to give us an idea of what to expect to come out of our function.

A function f from the set A (domain) to B (co-domain) will be denoted by $f : A \to B$. For every $a \in A$ there is exactly one $f(a) \in B$. Given an element a of the domain A, there is a process denoted by f the will allow us to find the unique $f(a)$ in the co-domain B. The abstractness of the term "process" is what we attempt to capture by the metaphor of the "black box." The blackness of the box indicates its opaque nature. It arises from the practice of electrical engineers using a box in a diagram to denote a procedure whose details are not yet germane to the discussion at hand.

A video to help clarify this is here.

3.2.3 Black Boxes

We will now examine how the black box metaphor can be used in visualizing processes such as function composition. First consider Figure 3.1 in which we have an isolated black box. Note that our black box has two whiskers on it: one on the left and one on the right. We read the diagram from left to right. The input, the element a from our domain, goes into our black box process f from the left; it is processed by f; then it exits on the right as $f(x)$.

There are times when it is more convenient to represent the input and/or output by more than one whisker. This occurs when the elements of the domain and/or co-domain are formed from one or more parts. Let us now illuminate this with a few well-chosen examples.

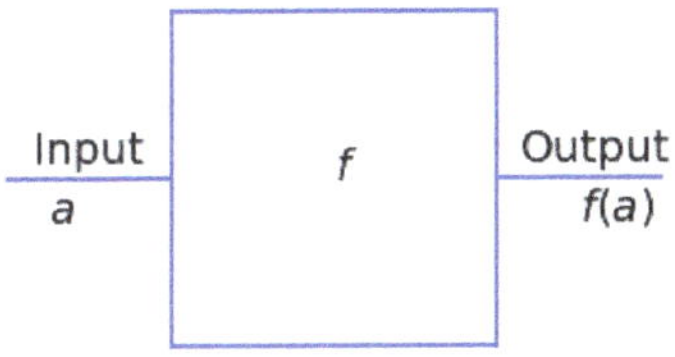

Figure 3.1: A Function as a Black Box

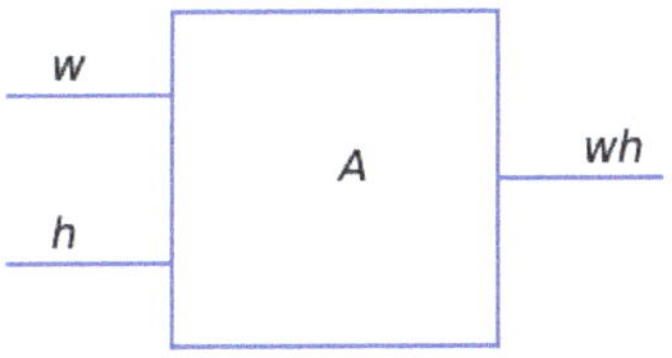

Figure 3.2: Area as a Black Box

A Black Box Example

Let use start out will a well-known function of two variables: Area. Typically, we will write area as a simple formula $A = wh$, where w denotes width and h denotes height. In order to illustrate our example, we will be more formal.

The domain D is the set of ordered pairs $\{(w, h) : w \geq 0, h \geq 0\}$ and the co-domain C is the interval $[0, \infty)$. The function for area can be given as $A : D \to C$ with $A(w, h) = wh$. We represent this by a black box in Figure 3.2

Now suppose that the dimensions vary according to time with the width being $3t$ units long at time t and the height being $5t$ units long at time t. This can be captured in the vector-valued function $D(t) = (3t, 5t)$ whose corresponding black box looks as in Figure 3.3 The composite function $(A \circ D)(t) = A(D(t))$ can be expressed in a diagram as in Figure 3.4.

A video to help with this concept is here.

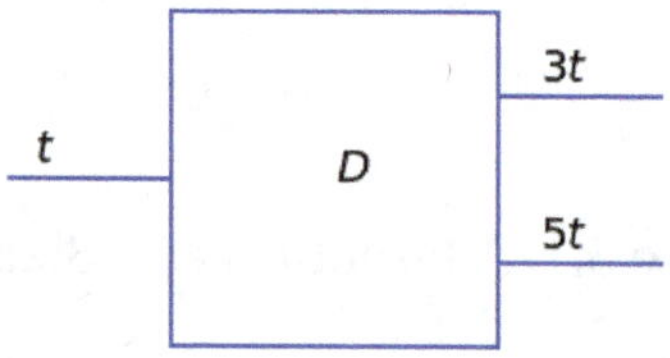

Figure 3.3: A Vector-valued function as a Black Box

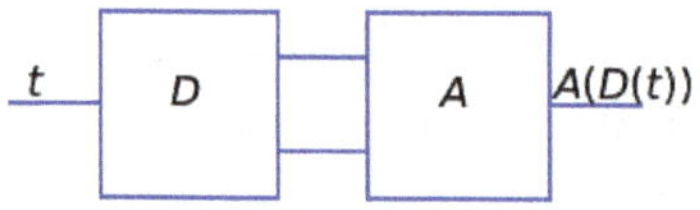

Figure 3.4: $A \circ D$ as a Black Box Diagram

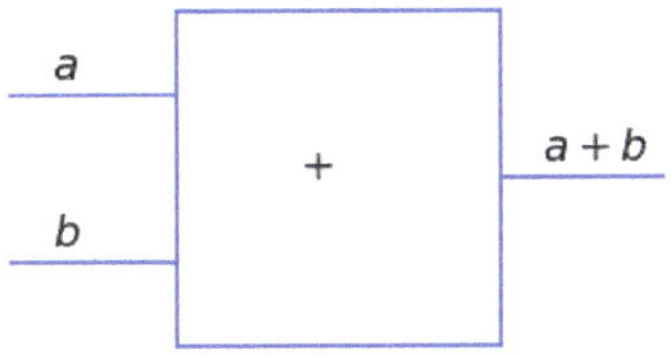

Figure 3.5: Addition as a Black Box Diagram

3.2.4 Binary Operations as Functions

While we are not yet accustomed to thinking of them this way, our four standard arithmetic operations ($+$, $-$, $*$, and $/$) are functions. To illustrate this, we will consider $+$, *i.e.* addition.

We will limit ourselves to the addition of natural numbers. In this case, the domain is the set of all ordered pairs (x, y) where x and y are both natural numbers. We will denote this set as $\mathbf{N}\times\mathbf{N}$. To use standard function notation, we will write $a+b$ as $+(a,b)$. That is to say we will denote the function by the symbol $+$ and our function will be $+:\mathbf{N}\times N \to \mathbf{N}$, with $+(a,b) = a+b$. As a black box we will write this as in Figure 3.5.

We can use this as a way to present complex arithmetical expressions in a visual way. For example, consider

$$\frac{x-y}{x+y}$$

as it is presented in Figure 3.6. Note that in the leftmost boxes x and y are split into the ordered pairs (vector-values) (x,x) and (y,y).

If you need come clarity try this video.

3.2.5 Boolean Functions

In this section, we will be interested in functions $f:\mathbf{Z}_2\times\mathbf{Z}_2\to\mathbf{Z}_2$, where $\mathbf{Z}_2\times\mathbf{Z}_2$ is the set of ordered pairs (x,y), where x and y are each elements of $\mathbf{Z}_2$. We can actually list the elements of the set out: $(0,0)$,$(0,1)$,$(1,0)$, and $(1,1)$.

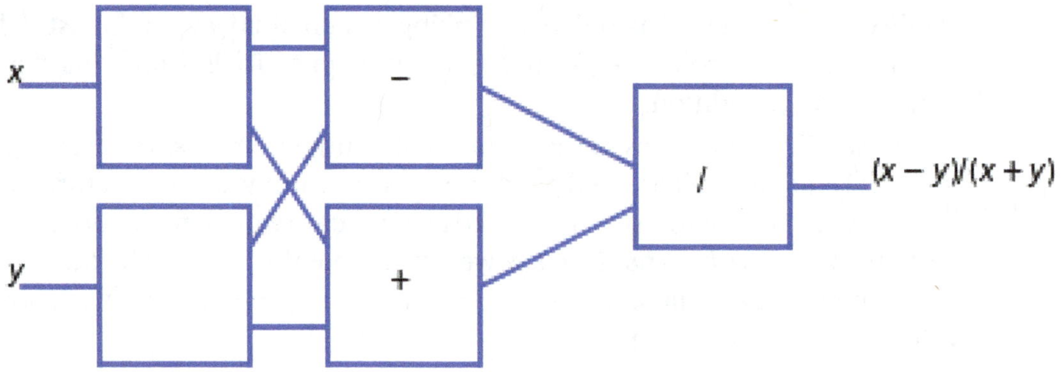

Figure 3.6: A Diagram for $(x-y)/(x+y)$

Let us begin with the function

$$< (x, y) = \begin{cases} 1 & \text{if } x < y \\ 0 & \text{all other } x, y \end{cases}. \tag{3.1}$$

Note that we are making the symbol “<” do double duty here. On the left is it is used to denote the function we are defining. On the right, it is use in its usual sense to say that x is less than y.

The interpreter or compiler for a computer language, does interpret the symbol $<$ as a function of two variables that is written in a strange way. In code like

$$\text{if } x < y \text{ then} \ldots$$

or

$$\text{if}(x < y)\{\ldots\}$$

The text $x < y$ is interpreted as the function $< (x, y)$.

Similarly, we can define a function $==: \mathbf{Z}_2 \times \mathbf{Z}_2 \to \mathbf{Z}_2$ to be

$$== (x, y) = \begin{cases} 1 & \text{if } x = y \\ 0 & \text{else} \end{cases} \tag{3.2}$$

In the Section3.3, we will apply this the so-called *logical operators.*

In a slightly different way, let U be a universal set and let $A \subset U$, we can think of “$\in$” as being a function $\in: U \to \mathbf{Z}_2$. We define

$$\in (x) = \begin{cases} 1 & \text{if } x \in A \\ 0 & \text{else} \end{cases}. \tag{3.3}$$

This is done using various notations in computer languages. For example, in Python, the keyword **in** is used in place of $\in$. There we write statements such as

$$\text{if } x \text{ in S then}$$

where “x in S” returns either **True** or **False**.

If this seems a little tricky, take a look at this.

x	y	$x \vee y$
0	0	0
0	1	1
1	0	1
1	1	1

Table 3.1: Logical OR

x	y	$x \wedge y$
0	0	0
0	1	0
1	0	0
1	1	1

Table 3.2: Logical AND

3.3 Logical Operators

Let x and y be Boolean numbers, *i.e.* elements of $\mathbf{Z}_2$. We define $x \vee y$ to be the maximum value of the set $\{x, y\}$; we pronounce $x \vee y$ as "x or y." Similarly, we define $x \wedge y$ to be the minimum of $\{x, y\}$, and pronounce $x \wedge y$ as "x or y." Another way to make this definition is to say $x \vee y$ is 1 if either x or y is equal to 1; $x \wedge y$ is equal to one if both x and y are equal to 1. These can be encoded in black box diagrams as with any binary operation, but this will done in a special way in Section 4.5.

We can also display how these logical operators work in Tables 3.1 and 3.2

The operators $\vee$ and $\wedge$ are both binary operators that operate on two variables[3]. We would also like to introduce an unitary[4] operator $\neg : \mathbf{Z}_2 \to \mathbf{Z}_2$. We call the operator $\neg$ not or NOT because it is contrary, *i.e.*

$$\neg(x) = \begin{cases} 1 & \text{if } x = 0 \\ 0 & \text{if } x = 1 \end{cases}.$$

Its truth table is easy which you can see for yourself in Table 3.3.

[3]Bi is Latin for two. If you didn't know that, now you do.

[4]Unus is Latin for one. You didn't know this was a language class, did you.

x	$\neg x$
0	1
1	0

Table 3.3: Logical NOT

x	y	$x \vee y$	$\neg(x \vee y)$	$\neg x$	$\neg y$	$\neg x \wedge \neg y$
0	0	0	1	1	1	1
0	1	1	0	1	0	0
1	0	1	0	1	1	0
1	1	1	0	0	0	0

Table 3.4: One of De Morgan's Laws

We will now use truth tables to explore the interaction of $\neg$, $\vee$, and $\wedge$. Consider Table 3.4. Note that the contents of the columns x and y go through all of the possible values that x and y can assume in combination. Observe that the rows of the columns $\neg(x \vee y)$ and $\neg x \wedge \neg y$ agree with each other. It follows from this that $\neg(x \vee y) = \neg x \wedge \neg y$. It can be shown similarly that $\neg(x \wedge y) = \neg x \vee \neg y$. We leave this to the reader. Together these have a name:

Proposition 3.3.1. *(De Morgan's Laws of Logic)*
For all $x, y \in \mathbf{Z}_2$,

1. $\neg(x \vee y) = \neg x \wedge \neg y$

2. $\neg(x \wedge y) = \neg x \vee \neg y$

In addition to De Morgan's Laws, we also have the Distributive Laws for Logical Operators:

Proposition 3.3.2. *(The Distributive Laws for Logical Operators)*
For $x, y, z \in \mathbf{Z}_2$,

1. $x \wedge (y \vee z) = (x \wedge y) \vee (x \wedge z)$

2. $x \vee (y \wedge z) = (x \vee y) \wedge (x \vee z)$

We will demonstrate the first of these in Table 3.5. The second will be left to the reader.

To see a short video on this, press on this.

x	y	z	$y \vee z$	$x \wedge (y \vee z)$	$x \wedge y$	$x \wedge z$	$(x \wedge y) \vee (x \wedge z)$
0	0	0	0	0	0	0	0
0	0	1	1	0	0	0	0
0	1	0	1	0	0	0	0
0	1	1	1	0	0	0	0
1	0	0	0	0	0	0	0
1	0	1	1	1	0	1	1
1	1	0	1	1	1	0	1
1	1	1	1	1	1	1	1

Table 3.5: Distributing $\wedge$ over $\vee$

3.4 Return to Sets

In this section we will return to the topic of sets. In particular, we will discuss some ways they interact with each other. Our approach differs from what is usually done as we will use methods that will construct the sets, *i.e.* methods that can be implemented in computer programs.

Given a universal set U, let A and B be subsets of U. We define $A \cup B$ to be the *union* of A and B, to be the set that contains all of the elements of A along with all of the elements of B. This is easy to understand intuitively as we are simply "dumping" the contents of the set together. We will now describe $A \cup B$ in two ways which are constructive in nature. The first is, perhaps, more intuitive, but the second is more concise and elegant and should be considered normative. See Figure 3.7. Note that each of these contains a loop that runs through all of the elements of U to test whether they should be a member of $A \cup B$. In order to insure this process stops, the set U must be finite. Method 1 is more intuitive as it will dump the sets together. By way of contrast, Method 2 uses logic. The Boolean function $(x \in A) \vee (x \in B)$ will return the value 1 (=**True**) for values of x that are in either A or in B. It is a smarter method.

With notation as above, we now define $A \cap B$, the intersection of A and B, to be the set of elements that A and B have in common. In Figure 3.8 we describe two ways to construct $A \cap B$. Method 1 is intuitive but potentially confusing, as the roles of A and B can be reversed, so there is no reason to put A first. Method 2, which uses logical operators to make the process more concise and elegant.

In addition to union and intersection, we will now define the difference of two sets. With the sets A and B as before, let $A - B$ be defined to be the set of

Method 1:

Input sets A, B
Put $C = \{\}$
For $x \in U$:
    If $x \in A$, then add x to C
    If $x \in B$, then add x to C
return C as $A \cup B$

or
Method 2:

Input sets A, B
Put $C = \{\}$
For $x \in U$:
    If $(x \in A) \vee (x \in B)$, then add x to C
return C as $A \cup B$.

Figure 3.7: Constructing the Union of A and B

the points of A that are not in B. Again we show a couple of ways to construct this set. See Figure 3.9

For the sake of clarity, you might consider watching this.

3.4.1 The Cardinality of Sets and Counting

Given the set S, the cardinality of a S is the number of elements it contains. In Section 3.1.3, we defined $\mathbf{N}$, the set of natural numbers. The set of cardinal numbers is the paradigm example of an infinite set. We say that its cardinality is *countably infinite*. As such, it falls beyond the scope of this course.

We will concentrate on finite sets. By definition, their elements can be counted. Our finite sets, $\mathbf{N}_n$, in Section 3.1.3 are paradigm examples of their type. In fact, we can make the definition that a *finite* set is one whose unique elements can be given completely in a sub-scripted list $e_1, e_2, \ldots, e_n$ for a fixed $n \in \mathbf{N}$. For example, $S = \{\text{Larry}, \text{Moe}, \text{Curly}\}$ has elements $e_1 = \text{Moe}$, $e_2 = \text{Larry}$, and $e_3 = \text{Curly}$.

Given a set containing distinct elements $e_1, e_2, \ldots, e_n$, we say the *cardinality*

Method 1

```
Input sets A, B
Put C = {}
For x ∈ U:
    If x ∈ A, then:
      If x ∈ B, then add x to C.
    else do nothing
return C as A ∩ B.
```

or

Method 2

```
Input sets A, B
Put C = {}
For x ∈ U:
    If (x ∈ A) ∧ (x ∈ B), then add x to C:
return C as A ∩ B.
```

Figure 3.8: Constructing the Intersection of A and B

Method 1:

Input sets A, B
Put $C = \{\}$
For $x \in U$:
    If $x \in A$, then:
        If $\neg(x \in B)$, then add x to C
    else do nothing
return C as $A - B$

or
Method 2:

Input sets A, B
Put $C = \{\}$
For $x \in U$:
    If $(x \in A) \wedge \neg(x \in B)$, then add x to C
return C as $A - B$

Figure 3.9: Constructing the $A - B$

of S is n and we write

$$|S| = n.$$

Though $|\cdot|$ also represents absolute value, this will have little opportunity to cause confusion.

Of course for every $n \in \mathbf{N}$, $|\mathbf{N}_n| = n$[5].

Our aim, now, is to derive some simple formulas for calculating the cardinality of finite sets. We will base these derivations on the following axiom:

Axiom 1. *If A and B are finite sets that don't have any common elements,* i.e. *$A \cap B = \emptyset$, then*

$$|A \cup B| = |A| + |B|.$$

This is stated as an axiom–even though it might be provable in the right context–because it is so intuitive. We will now use it to derive a common formula for $|A \cup B|$ for the case where A and B might have elements in common.

Consider that

$$A = (A - B) \cup (A \cap B).$$

The sets $A - B$ and $A \cap B$ have no elements in common. It follows that

$$|A| = |A - B| + |A \cap B|.$$

Similarly,

$$|B| = |B - A| + |A \cap B|$$

Now $A \cup B = (A - B) \cup (B - A) \cup (A \cap B)$. Since $A - B$, $B - A$, and $A \cap B$ don't have any elements in common,

$$\begin{aligned} |A \cup B| &= |A - B| + |B - A| + |A \cap B| \\ &= (|A - B| + |A \cap B|) + (|B - A| + |A \cap B|) - |A \cap B| \\ &= |A| + |B| - |A \cap B| \end{aligned}$$

with the final step coming from substituting the two previous equations. Therefore,

Proposition 3.4.1. *For finite sets A and B,*

$$|A \cup B| = |A| + |B| - |A \cap B|.$$

To help with your understanding, you can watch this.

[5] For those of you who are curious, $|\mathbf{N}| = \aleph_0$. This is one of the so-called *transfinite* cardinals.

3.5 Combinations and Permutations

3.5.1 An Introduction to Combinations

Let us consider the set $S = \{1,2,3\}$. We will list all of the subsets of S as follows:

$$\begin{aligned}&\emptyset\\&\{1\},\ \{2\},\{3\}\\&\{1,2\},\ \{1,3\},\ \{2,3\}\\&S\end{aligned}$$

The subsets have been listed in a particular way. The lone set on the top line has zero elements. The three sets on the next line each contain exactly on element. The three sets in the next line each contain two elements. The lone set on the bottom contains exactly three elements.

Another way to say this is that $\emptyset$ is the only subset we can make from S by choosing 0 elements; $\{1\}$, $\{2\}$, and $\{3\}$ are the three subsets we can make from S by choosing 1 element; $\{1,2\}$, $\{1,3\}$, and $\{2,3\}$ are the subsets we can make from S by choosing 2 elements. Finally, S is the only subset of S we can make by choosing 3 elements.

In this part of the course, we are concerning ourselves with counting subsets in this second way. We will consider a set S that contains $n \geq 0$ elements. We want to know how many subsets there are of size $0 \leq r \leq n$. We will refer to this number as *n choose r* and will denote it as ${}_nC_r$. [6]

In our example above ${}_3C_0 = 1$, ${}_3C_1 = 3$, ${}_3C_2 = 3$, and ${}_3C_3 = 1$. Our aim is to derive a formula for ${}_nC_r$. We will have to seek up on this, but first we will establish a connection between combinations and polynomials.

3.5.2 Combinations and Polynomials

To see the connection between combinations and polynomials, consider the product of binomials $(1 + a_1x)(1 + a_2x)(1 + a_3x)$. Without too much trouble, we can expand it as follows:

[6]This is also frequently denoted by $\binom{n}{r}$. We use ${}_nC_r$ because it connects to the notation ${}_nP_r$ that we will be using later.

$$\begin{aligned}(1+a_1x)(1+a_2x)(1+a_3x) &= 1\\ &+ (a_1+a_2+a_3)x\\ &+ (a_1a_2+a_1a_3+a_2a_3)x^2\\ &+ a_1a_2a_3x^3.\end{aligned}$$

When we compare this to the example of Section 3.5.1, we see a correspondence between the of coefficient of the x^r term and the subsets of $S = \{1,2,3\}$ of size r. The coefficient of x^3 is $a_1a_2a_3$. The indices of this product are 1,2,and 3, and the only subset of S of size 3 is $\{1,2,3\}$. The coefficient of x^2 is $(a_1a_2 + a_1a_3 + a_2a_3)$, which has three terms. These clearly correspond to the subsets $\{1,2\}$, $\{1,3\}$, and $\{2,3\}$, which are the subsets of S of size 2. The coefficient of x is $a_1 + a_2 + a_3$, which have three terms corresponding to the sets 1, 2, and 3. Finally, the coefficient of x^0 is 1, which corresponds to the empty set.

We may apply this in a number of ways. First of all, let us consider the subsets of $\{1,2,3,4\}$ which we will list below:

$$\begin{array}{l}\emptyset\\ \{1\}, \{2\}, \{3\}, \{4\}\\ \{1,2\}, \{1,3\}, \{1,4\}, \{2,3\}, \{2,4\}, \{3,4\}\\ \{1,2,3\}, \{1,2,4\}, \{1,3,4\}, \{2,3,4\}\\ \{1,2,3,4\}\end{array}$$

This table will allow us to expand the product of binomials given below:

$$\begin{aligned}(1+a_1x)(1+a_2x)(1+a_3x)(1+a_4) &= 1\\ &+ (a_1+a_2+a_3+a_4)x\\ &+ (a_1a_2+a_1a_3+a_1a_4+a_2a_3+a_2a_4+a_3a_4)x^2\\ &+ (a_1a_2a_3+a_1a_2a_4+a_1a_3a_4+a_2a_3a_4)x^3\\ &+ a_1a_2a_3a_4x^4.\end{aligned}$$

Indeed, for any n, the coefficient of x^r in $(1+a_1x)(1+a_2x)\cdots(11a_nx)$ is the sum of all possible products of $a_{i_1}\cdots a_{i_r}$ where $\{i_1,\ldots,i_r\}$ is a subset of size r chosen from $\{1,\ldots,n\}$.

A second application of this connection comes when we consider the special case $a_1 = a_2 = \cdots = a_n = 1$. That is to say $(1+a_1x)\cdots(1+a_nx) = (1+x)^n$.

Then, for example,

$$\begin{aligned}(1+x)^3 &= 1\\ &+ 3x\\ &+ 3x^2\\ &+ x^3.\end{aligned}$$

Note that the coefficient of x^r in this is equal to ${}_3C_r$. Again, this is not a coincidence. In general, we can put this in the following named formula:

Proposition 3.5.1. [7]

$$(1+x)^n = \sum_{r=0}^{n} {}_nC_r x^r.$$

We can look at our example of the expansion of $(1+a_1x)(1+a_2x)(1+a_3x)(1+a_4)$ and in letting $a_1 = a_2 = a_3 = a_4 = 1$ see that ${}_4C_0 = 1$, ${}_4C_1 = 4$, ${}_4C_2 = 6$, ${}_4C_3 = 4$, and ${}_4C_4 = 1$.

It is always going to be the case that ${}_nC_0 =_n C_n = 1$. Look at our examples to see why.

3.5.3 Lists

We all know what lists are. We've made to do lists and shopping lists. Dealing with them in mathematics and computer science will not be difficult, but it will require us to be precise. Let us begin.

Let I be an *indexing set.* That is to say that I is a set of consecutive integers, usually $\{1,2,\ldots,n\}$ or $\{0,1,,\ldots,n-1\}$, where n is a positive integer. Let V be a set of values. These values can be number or names or more abstract entities. A *list* of n values from V is a function $e: I \to V$.

As said at the beginning, we typically do deal with lists at this level of formality. Given and indexing set $I = \{1,2,\ldots,n\}$ indicated above as $e_1,\ldots,e_n$. To be even more concrete, $2,1,3,4,2,1$ would be a list of values from the set $\{1,2,3,4\}$ with an indexing set that has 6 elements.

The key property to focus on with lists is **order**. Except for the last item in a list, each item precedes another; except for the first item in the list, each item follows another.

[7] This is more commonly written as $(1+x)^n = \sum_{r=0}^{n} \binom{n}{r} x^r$.

Lists are common data structures in programming languages. Often they are referred to as vectors or arrays, but we will stick to the generic term list for the name. Each programming language will have its own syntax to use in dealing with lists. For example, in Python the specific list we give above would be given as

$$e = [2, 1, 3, 4, 2, 1].$$

In Python, the indexing set for e would be $\{0, 1, 2, 3, 4, 5\}$. We could access the particular elements as follows: $e[0] = 2$, $e[1] = 1$, $e[2] = 3$, $e[3] = 4$, $e[4] = 2$ and $e[5] = 1$.

3.5.4 Counting Lists

In this section, we will talk about counting the number of lists r items long formed from a set of n fixed values. This is the same question as how many functions are there from the set $\{1, 2, \ldots, r\}$ to the set $\{1, 2, \ldots, n\}$. We present the answer in the form of the following proposition.

Proposition 3.5.2. *The total number of lists of length r from a set of n elements is*

$$n^r.$$

As order is the key property of lists, we can model a list as a sequence of boxes that are connected to each other in a fixed order as in Figure 3.10.

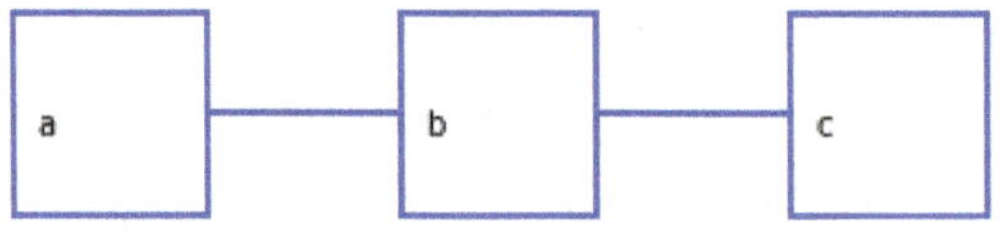

Figure 3.10: Model of a List

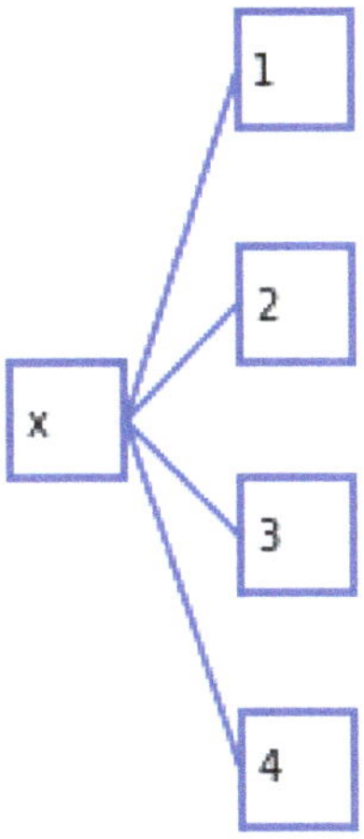

Figure 3.11: Possibilities for following the value x

We imagine each of these boxes containing an element of the value set. To be definite, suppose our value set is $V = \{1,2,3,4\}$ and that x is the value in a box that has a successor. The various ways the next box can be filled can be modeled as in Figure 3.11 Suppose that $I = \{1,2,3\}$ is the indexing set. We can model this situation as in Figure 3.12 As we vary x through 1, 2, 3, and 4, we can see there are $4 \times 4 \times 4 = 4^3 = 64$ different lists from the value set $V = \{1,2,3,4\}$.

3.5.5 Permutations

The word *permutation* is less common than the word *list*, but nevertheless it still occurs in common parlance. To permute objects means to put them into a different order. In general, we can say a *permutation* is a list of items in which no item is repeated.

For those who want to see a mathematically precise definition let I be an indexing set and let V be a set of values. Then we say $p : I \to V$ is a *permutation* of V if $p(n) \neq p(m)$ whenever $n \neq m$. In this book, we will not need this level of formality.

The property of permutations that separate them from generic lists is that they don't repeat values. This cannot be repeated often enough.

We will now count permutations as we did lists. We will begin by consider-

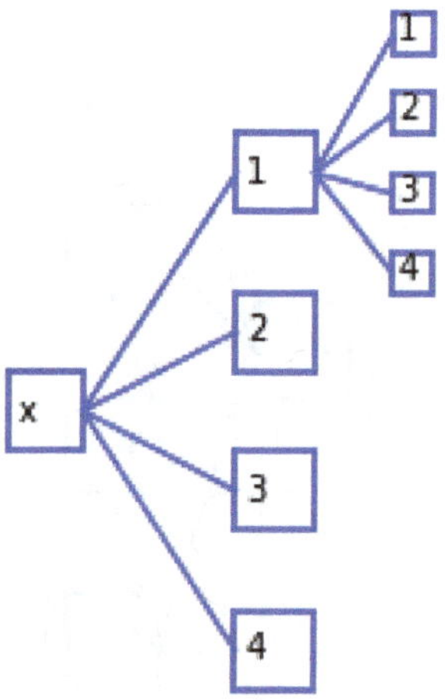

Figure 3.12: Possibilities in a List of Length 3

ing a special case. We will count the number of permutations of the indexing set $\{1,2,\ldots,r\}$.

Proposition 3.5.3. *The number of permutations of the set* $\{1,2,\ldots,r\}$ *is equal to* $r!$.

We can see how we arrive at $r!$ by using the box diagrams we used to count lists. We will do this for the specific index set $\{1,2,3,4\}$. Say we begin with the value 1. Then we can list the possibilities in Figure 3.13. We can actually count that there are 6 permutations that begin with 1. As there will be 6 permutations for each of the values 1,2,3, and 4, there will be a total of $4\times 6 = 24 = 4!$ permutations altogether.

Let us now move to the more general case, where we don't necessarily permute all of the items of the indexing set. That is to say we consider permutations of r values from a set of $n \geq r$ values.

To aid exposition, let $n = 4$ and $r = 2$. For a permutation beginning with 1, the possibilities are as in Figure 3.14. There are 3 permutations that start with 1. As we have 3 possibilities that follow each of the four starting values, there are $4\times 3 = 12$ permutations of 2 elements taken from the set of four values.

The number of permutations of r values taken from a set of n values is a product. The first factor is n, the seconds is $n-1$, and so on reducing by 1

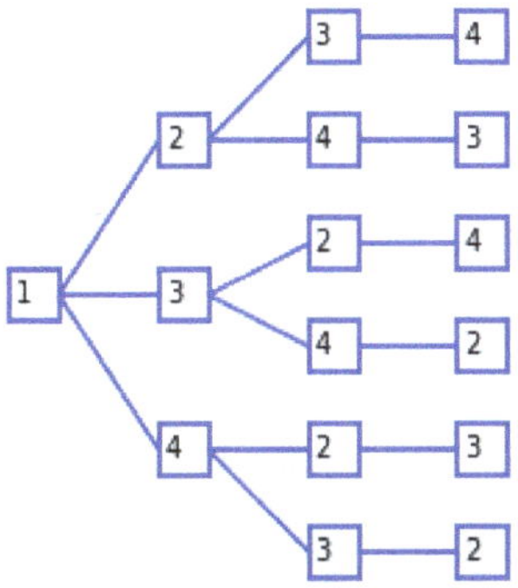

Figure 3.13: Box Model for 4!

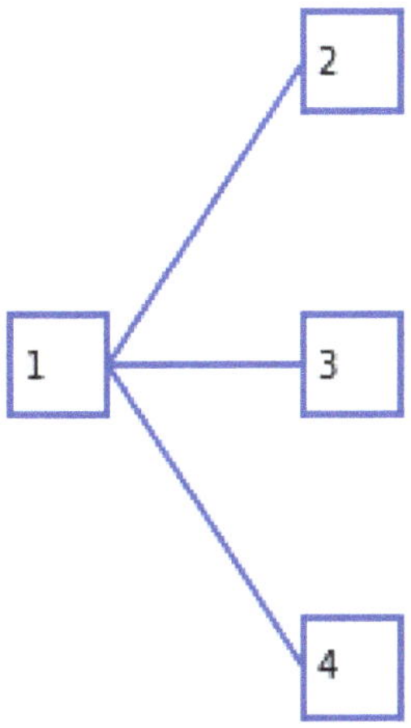

Figure 3.14: Possible Next Choices

until we have a total of r factors. This can be written in a comparatively simple formula which is given the the following proposition.

Proposition 3.5.4. *The number of permutuations of r items taken from a set of n values is given by*

$$_nP_r = \frac{n!}{(n-r)!}.$$

3.5.6 Connecting Permutations and Combinations

We now come to a point where we can make the connection between permutations and combinations. It involves finding a different way to calculate the number of permutations of r elements from a set of n elements.

To begin, let V be a set of n values. Now choose a subset S of V that contains r values. As S has r elements, there are $r!$ ways to order it. Given that there are $_nC_r$ ways to choose the subset S, we can calculate that there are

$$_nC_r \cdot r!$$

permutations of r elements of V, i.e.

$$_nP_r =_n C_r \cdot r!.$$

Rearranging this yields

$$_nC_r = \frac{_nP_r}{r!}$$

as

$$_nP_r = \frac{n!}{(n-r)!}$$

we can calculate

$$_nC_r = \frac{1}{r!} \cdot \frac{n!}{(n-r)!}$$

that is to say,

$$_nC_r = \frac{n!}{r!(n-r)!}. \tag{3.4}$$

3.6 Exercises for Chapter 3

1. Let $U = \{1,2,3\}$.
   (a) List all of the subsets of U that contain 1.
   (b) List all of the nonempty subsets of U that don't contain 1.
   (c) List all of the subsets of U.
   (d) How many subsets of U are there?

2. Let $U = \{1,2,3,4\}$.
   (a) List all of the subsets of U that contain 1.
   (b) List all of the nonempty subsets of U that don't contain 1.
   (c) List all of the subsets of U.
   (d) How many subsets of U are there?

3. Write out the elements of the sets described below explicitly.
   (a) $\{n \in \mathbf{N}_{15} : 3|n\}$
   (b) $\{n \in \mathbf{N}_{15} : 5|n\}$
   (c) $\{n \in \mathbf{N}_{15} : 2|n\}$

4. Write out the elements of the sets described below.
   (a) $\{n \in \mathbf{N} : n^2 + 1 < 10\}$
   (b) $\{n \in \mathbf{N} : 20 - 3n^2 > 0\}$

5. Write out the elements of the sets described below.
   (a) $\{x \in \mathbf{Z}_2 : x^2 - 1 \equiv_2 0\}$
   (b) $\{x \in \mathbf{Z}_2 : x^2 + x + 1 \equiv_2 0\}$

6. Give the domains of each of the following functions.
   (a) $exp : (-\infty, \infty) \to (0, \infty)$
   (b) $\sqrt{\ } : [0, \infty) \to [0, \infty)$
   (c) $\sqrt{4 - x}$

(d) $\sqrt{4-x^2}$

7. Give the (maximal)co-domains of each of the following functions.

   (a) $exp: (-\infty,\infty) \to (0,\infty)$

   (b) $\sqrt{\ }: [0,\infty) \to [0,\infty)$

   (c) $\sqrt{4-x}$

   (d) $\sqrt{4-x^2}$

8. Look at Figure 3.2. Using it as a model along with the fact that volume can be calculated as $V = whl$ where w is width, h is height, and l is length, do the following.

   (a) Write the formula for volume using functional notation.

   (b) Give a black box representation of the function for volume.

9. Let $D: [0,\infty) \to [0,\infty) \times [0,\infty) \times [0,\infty)$ where $[0,\infty) \times [0,\infty) \times [0,\infty)$ is the set of ordered triples (r, s, t) where $r, s, t \in [0,\infty)$. Say $D(t) = (3t, 5t, 7t)$. Using Figure 3.3 as a model, draw a black box for D.

10. Using Figure 3.4 as well as Exercises 8 and 9 to create a black box diagram for $V \circ D$.

11. Using Figure 3.6 as a model, create black box diagrams for the following expressions.

    (a) $x(y+z)$

    (b) $xy+z$

    (c) $x/y+z$

    (d) $x+yz$

    (e) $x/(x+y)$

12. Perform the following calculations.

    (a) $<(0,0)$

    (b) $<(0,1)$

    (c) $<(1,0)$

(d) $< (1,1)$

13. Put the results from Exercise 12 in a truth table.

14. Perform the following calculations.

    (a) $== (0,0)$

    (b) $== (0,1)$

    (c) $== (1,0)$

    (d) $== (1,1)$

15. Put the results from Exercise 14 in a truth table.

16. Using Table 3.4 as a model, prove $\neg(x \wedge y) = \neg x \vee \neg y$.

17. Using Table 3.5 as a model, prove $x \vee (y \wedge z) = (x \vee y) \wedge (x \vee z)$.

18. Without using truth tables, use the formulas of this section to show

$$\neg(\neg x \vee y) = x \wedge \neg y.$$

19. Consider $\neg x \vee y$. Use a truth table to show that this expression is always equal to y when $x = 1$.

20. Given $U = \mathbf{N}$, $A = \{1,6,7,8\}$ and $B = \{2,3,4,8,9\}$, find the following.

    (a) $A \cup B$

    (b) $A \cap B$

    (c) $A - B$

21. Given $U = \mathbf{N}$, $A = \{m \in U | m = 2n \text{ for } n \in \mathbf{N}\}$ and $B = \{m \in U | m = 2n - 1 \text{ for } n \in \mathbf{N}\}$, find the following.

    (a) $A \cup B$

    (b) $A \cap B$

    (c) $A - B$

22. If $A \subset B$, answer the following.

    (a) What is $A - B$?

(b) What is $A \cap B$?

(c) What is $A \cup B$?

23. Let $U = \mathbf{N}_9$. Let $A = \{3,6,9\}$ and let $B = \{2,4,6,8\}$. For each $x \in U$, calculate the following:

    (a) $(x \in A) \wedge \neg(x \in B)$

    (b) $(x \in A) \wedge (x \in B)$

    (c) $(x \in A) \vee (x \in B)$.

24. Let $U = \mathbf{N}_9$. Let $A = \{3,6,9\}$ and let $B = \{2,4,6,8\}$. As appropriate, use the pseudo-code from the Method 2's given to preform the following calculations. Number the steps and record the contents of C at the end of each step.

    (a) $A \cup B$

    (b) $A \cap B$

    (c) $A - B$

25. Let $A = \{n \in N_{1000} : 3|n\}$ and let $B = \{n \in N_{1000} : 5|n\}$. Find the following quantities.

    (a) $|A|$

    (b) $|B|$

    (c) $|A \cap B|$

    (d) $|A \cup B|$

26. Let $A = \{n \in N_{1000} : 8|n\}$ and let $B = \{n \in N_{1000} : 2|n\}$. Find the following quantities.

    (a) $|A|$

    (b) $|B|$

    (c) $|A \cap B|$

    (d) $|A \cup B|$

27. List all of the subsets of each of the following. Do so in a stratified way, with the first row c containing the empty set, the second with all of the sets with one element, and so forth.

(a) $\{1,2\}$

(b) $\{1,2,3,4\}$

(c) $\{1,2,3,4,5\}$

28. Refer back to Problem 27 in order to answer each of the following.

(a) ${}_2C_1 =$

(b) ${}_4C_2 =$

(c) ${}_4C_3 =$

(d) ${}_5C_3 =$

(e) ${}_5C_4 =$

29. Expand the following polynomials.

(a) $(1+x)(1+2x)(1+3x)$

(b) $(1+2x)(1+5x)(1+7x)$

(c) $(1+2x)(1+3x)(1+4x)(1+5x)$

(d) $(1+x)^5$

30. How many lists of length 7 can be made from the set $\{1,2\}$

31. Perform the following calculations.

(a) 8!

(b) ${}_8P_2$

(c) ${}_{52}P_7$

32. Calculate each of the following.

(a) ${}_{20}C_{10} =$

(b) ${}_{44}C_{22} =$

(c) ${}_{44}C_{23} =$

(d) ${}_{45}C_{22} =$

(e) ${}_{45}C_{23} =$

Chapter 4

Numbers and Numerals

4.1 Introduction

Back in the early 1970s when I first began to struggle with the basics of arithmetic, I was introduced to something called "The New Math." The New Math was well-meant like all of the subsequent efforts to improve the instruction of mathematics have been. It attempted many things, but the thing I am thinking of now is that it made a big deal about separating the concepts of "numeral" and number. A numeral is a means of denoting a number, but a number is the thing itself.

At the time, being about 10 years old, I thought this was stupid, and to be sure I do so now, but for different reasons.

At the time I thought that I already knew the difference between a numeral and a number. Now I am no longer sure I know what a number is.

In those days, they segued from teaching this distinction to teaching us about Roman Numerals. They might've also taught us about putting numbers into different bases, but I cannot recall.

My reason for bringing this all up now is that the purpose of this chapter is to show–to a certain degree of precision–how computers add numbers. The first step is to represent numbers in a way they can be encoded into the computer, i.e. in binary. The next step is to describe the addition algorithm on a macro level. The final step, which will be quite protracted, will be to describe how this can be implemented electronically via half-adders, full-adders, logic gates, and mathematically simplified transistors.

This will be done from the point of view of mathematics, not electrical engineering, but giving due respect to the electrical engineers. May God bless them.

4.2 Number Representation

In §4.1, I spoke about the distinction between numbers and numerals. I am interested in representing numbers in a way computers can deal with them. In a certain sense, the most basic idea of a number is that something is either there or it is not; there is either something or there is nothing; something is either True or it is False. When there is nothing we can denote it as 0; when there is something we can denote it as 1. When call this a *boolean* number.

That first step was a little philosophical and I apologize for that. The next step is more down to earth: counting. When we count, something even little children can do, we create the counting numbers which mathematicians refer to as *positive integers* or *natural numbers*: $1, 2, 3, 4, \ldots$. We are giving names to positions in an order. You can think of laying out coins or peanuts or corn nuts or a mixture of all of them while giving them names.

Over the ages, we've learned to be systematic in assigning the names so that we always know how to write down the next number even if we don't know how to pronounce it. We do this by using the so-called Hindu-Arabic numbers which we more generically refer to as base-ten numbers or decimal numbers. (Note how quickly we become careless about the numeral number distinction in ordinary speech.)

In working with computers, we use base-two numbers which are better known as *binary* numbers: 1, 10, 11, 100 ... Note that we can go on naming our numbers forever while we very quickly lose the ability to read them in their common names.

It now becomes useful for us to write this in the language of mathematics. When we have a positive integer b, there is an integer $n \geq 0$ and a sequence of boolean numbers $b_0, b_1, \ldots, b_n$ such that

$$b = 2^n \cdot b_n + \cdots + 2 \cdot b_1 + b_0.$$

For the sake of brevity, we would typically suppress the 2^i in the expansion and write this number as $b = (b_n b_{n-1} \ldots b_1 b_0)_{\mathrm{bin}}$. At times when no confusion is possible, we will write $b = b_n b_{n-1} \ldots b_1 b_0$.

To help you get started, do the exercise associated with this video.

4.3 Numbers in a Computer

As mentioned before, the New Math made a big deal about separating the concepts of numeral and number. I am now going to make a big deal out of separating a number and its encoding in a computer. Representing the number in binary is just the first step to encoding it. You may have read somewhere that we use binary numbers because 1 denotes a switch being on and 0 represents a switch being off. That is not so much wrong as it is incomplete. Having a 1 can also mean that the voltage is at or above a certain threshhold while having a 0 means that voltage is below that threshhold. It can also mean other things. Viewed on paper it is an abstraction. Abstract thinking is wonderful, but we do need to keep in mind that our abstractions refer to physical entities. We live in a physical world.

Speaking of living in a physical world, we are limited to the size of number that we can encode. When we represent a number in binary as $b = b_n b_{n-1} \dots b_1 b_0$, each b_i is referred to as a *bit*. The number b have have just named has $n+1$ bits. At the time of this writing, computers which can handle numbers 64-bit numbers are quite common.

In this treatment, we will generically talk about N-bit numbers and in our examples we will set $N = 4$[1]. This will limit us to using only the numbers between 0 and $2^4 - 1 = 15$, inclusive, in examples but this will not hinder the process we are learning about.

In the language of computer science, we will be dealing with 4-bit unsigned integers. Dealing with signed integers is only slightly more complicated, but will be deferred to another time. We will now list all of the 4-bit unsigned integers in Table 4.3.

For some extra clarity, take a look at this video.

4.4 Binary Addition

Though most of you already know how to do binary addition by hand or could at least take a creditable stab at it, I would like to start this section with a simple example. Consider 11 plus 5 being represented in binary as

$$\begin{array}{ccccc} & 1 & 0 & 1 & 1 \\ + & 0 & 1 & 0 & 1 \end{array}$$

[1]Using $N = 64$ would add nothing to our understanding and make our exposition impossible to read. The fact that it would make the writer's life hell is just a coincidence.

b	$(b)_{\mathrm{bin}}$
0	0000
1	0001
2	0010
3	0011
4	0100
5	0101
6	0110
7	0111
8	1000
9	1001
10	1010
11	1011
12	1100
13	1101
14	1110
15	1111

Table 4.1: 4-bit unsigned integers

We begin the addition algorithm by adding the rightmost bits. In this case $1+1=10$. In our process we will write down the number $s=0$ (s stands for sum) and $c=1$ (c stands for carry). This will appear as

$$\begin{array}{cccccc} & & 1 & 0 & 1^1 & 1 \\ & + & 0 & 1 & 0 & 1 \\ \hline & & & & & 0 \end{array}$$

Let us note that the first step requires only two inputs: one bit from each of the summands. There are two outputs: sum and carry.

In the next step, we go to the rightmost but one column. We take the carry from the previous step, which you recall was 1 in this case, and add it to the 1 from the top summand. This gives us $1+1=10$. We now add the 0 from the lower summand this gives us $10+0=10$. So we again put down an $s=0$ and carry $c=1$. This will appear as

$$\begin{array}{cccccc} & & 1 & 0^1 & 1^1 & 1 \\ & + & 0 & 1 & 0 & 1 \\ \hline & & & & 0 & 0 \end{array}$$

What we need to know from this step is that there are actually three inputs: one bit from each summand as before and the carry value c from the previous step.

Continuing in this way until we are done will give us

$$\begin{array}{cccccc} & & 1^1 & 0^1 & 1^1 & 1 \\ & + & 0 & 1 & 0 & 1 \\ \hline & & & 0 & 0 & 0 \end{array}$$

and

$$\begin{array}{cccccc} & & 1^1 & 0^1 & 1^1 & 1 \\ & + & 0 & 1 & 0 & 1 \\ \hline & 1 & 0 & 0 & 0 & 0 \end{array}$$

The final sum of 11 and 5 is 16. This would result in an overflow error on a 4-bit machine.

In the sequel, we will ape this process and create two black box devices: the half-adder and the full-adder. We will then shed light on what goes on inside these black boxes. This will take a while.

In the meantime, you might want to watch this.

Inputs			Outputs	
b_1	b_2		s	c_{out}
0	0	→	0	0
0	1	→	1	0
1	0	→	1	0
1	1	→	0	1

Table 4.2: Half-Adder

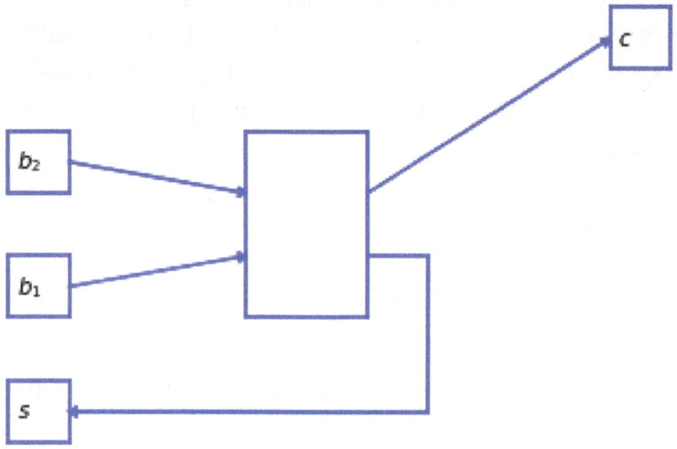

Figure 4.1: Half-Adder

4.4.1 The Half-Adder

The *half-adder* is an electronic device[2] that will take in two bits and return their sum in the form of two bits called the sum (denote by s) and the carry (denoted by c). How it works is illustrated in Table 4.2.

The table describes how the half-adder works in a mathematical way. This is fine, as far as it goes, but when time comes to design a machine we will need to think the half-adder as being a chip that we will plug in to a printed circuit board(PCB). To aid in this, we have created the diagram in Figure 4.1. In the figure we have inputs labeled b_1 and b_2. These will have current or voltage (or

[2]Think of it as a chip.

Inputs				Outputs	
c_{in}	b_1	b_2		s	c_{out}
0	0	0	→	0	0
0	0	1	→	1	0
0	1	0	→	1	0
0	1	1	→	0	1
1	0	0	→	1	0
1	0	1	→	0	1
1	1	0	→	0	1
1	1	1	→	1	1

Table 4.3: Full-Adder

the lack thereof) associated with them according to whether they are 1s or 0s. There is an outflow of current or voltage to s and c.

The phrase "black box" is meant to evoke mystery. We know what the black box does, but we do not know how. Our aim is to dispel this mystery to a certain degree. We will push to the boundary of physics, but we will not step over.

For help getting started with the exercises on this, look here.

4.4.2 Full-Adder

We will present a *full-adder* in a black box in a way similar to what we did for the half-adder above. See Table 4.3 Note that we have three inputs the bits b_1 and b_2 from our summands and the carry from our previous step that we denote as c_{in}. We have outputs: our sum s and our carry that we denote as c_{out}.

As before, we will want to think of this as a chip that will be plugged into a PCB. We do this in Figure 4.2. Again the lines are standing in for electrical connections, with c, b_1, and b_2 denoting inputs from the bit carried in and the bits from the summands. The outputs c and s denote the bit we carry out and the sum that we write down at this step, respectively.

We wish to justify the terminology half-adder and full-adder by showing that the full-adder is composed of two half-adders. This is, in fact, "almost" true. We need one more item. This appears to be a small thing at first, but it will open the world of logic gates to us.

If you need a little extra help, try this.

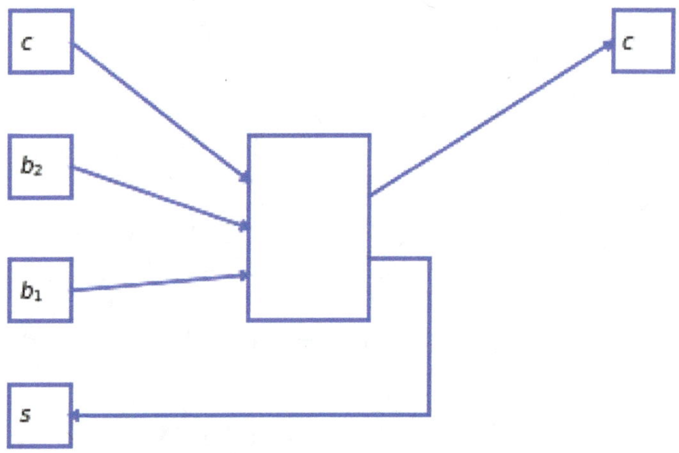

Figure 4.2: Full-Adder

4.5 Our First Taste of Logic

In § 4.2, we mentioned boolean numbers. there we said we could use 1 to denote if something was present or a statement was true and 0 if a statement was false or something was absent. This is of use in Boolean Algebra and logic. These are broad topics whos surface we will only begin to skim. In order to get our toes wet, while finding a way to demystify the full-adder, we will now introduce *logical OR.*

The logical OR operation is an binary operation on bits. In this context, binary means it has two operands like the plus operation does. We denote logical OR by the symbol $\vee$. Let p and q be boolean values or bits. Then $p \vee q$ (read as p or q) is 0 exactly when both p and q are 0 and is equal to 1 all other times. We make this specific in Table 4.4.

As with all of our operation, we are interested in "creating a chip" to implement this operation and in plugging that chip into a PCB. As logical OR is a basic operation, there is a standard wiring diagram symbol for it. We call the symbol an *OR-gate,* and it is shown in Figure 4.3.

Note that the OR-gate is shaped vaguely like an arrowhead. We can think of the arrowhead as indicating the direction of motion of current through it. The two horizontal segments on the lefthand side are inputs where the electrical

Inputs		Output
b_1	b_2	$b_1 \vee b_2$
0	0	0
0	1	1
1	0	1
1	1	1

Table 4.4: Logical OR

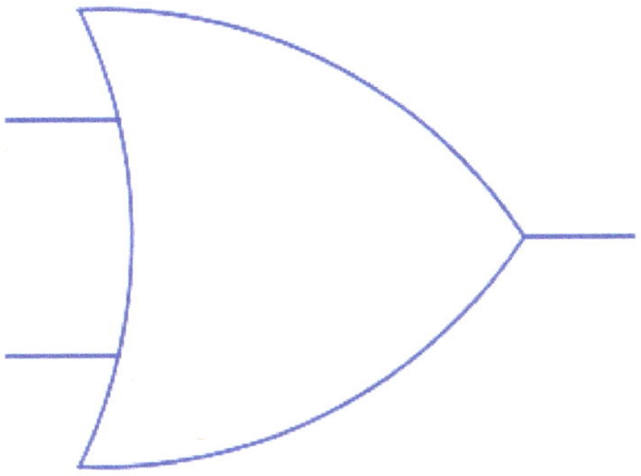

Figure 4.3: An OR-gate

manifestations of p and q are connected. Let horizontal segment on the right is where the current associated with $p \vee q$ is output.

This is just the first of many logic gates. To see how they may be combined, have a look here.

4.6 Full-Adder, Revisited

Let us now revisit the process of adding binary numbers. In every step but the first one, we have three inputs: c, carry-in; b_1, a bit from the first summand; b_2, a bit from the second summand. While using the full-adder to do this creates the impression that this is done in one seemless step, to anyone doing this by hand, it is absolutely clear that we are combining two operations.

First we perform the addition $c + b_1$ on a half-adder, call it HA_1. This will yield a carry-out bit that we will call c_1 and a sum bit that we will call s_1.

After doing this we perform the second operation on another half-adder that we will call HA_2. this will yield a carry-out of c_2 and a sum of s. This s is the sum of the full-adder.

The carry-out of the full-adder is obtained by combining c_1 and $c2$. This could be done simply by adding them together, and indeed, this is what we do when we do the work by hand. However, our purpose here is to try to get something simpler.

Let us first note that *at most* one of the c_1 and c_2 is equal to 1. If they are both 0, then $c_1 + c_2 = 0$ and $c_1 \vee c_2 = 0$. If exactly one of c_1 and c_2, then $c_1 + c_2 = 1$ and $c_1 \vee c_2 = 1$. Therefore, the final carry-out of our full-adder is $c_1 \vee c_2$. We can therefore diagram a full-adder in term of half-adders as in Figure 4.4.

We are still in a position where we've put something we don't know how to design in terms of transistors into other things we don't know how to design in terms of transistors, but at least we've dealt with the full-adder. To deal with the half-adder and the OR-gate we will have to learn some more logic.

4.7 Other Logical Operators

In this section, we will learn about two more basic logical operators in addition to OR: AND and NOT. Besides these, we will learn about the composite operators NAND and exclusive OR (XOR). We will also learn about the corresponding logic gates.

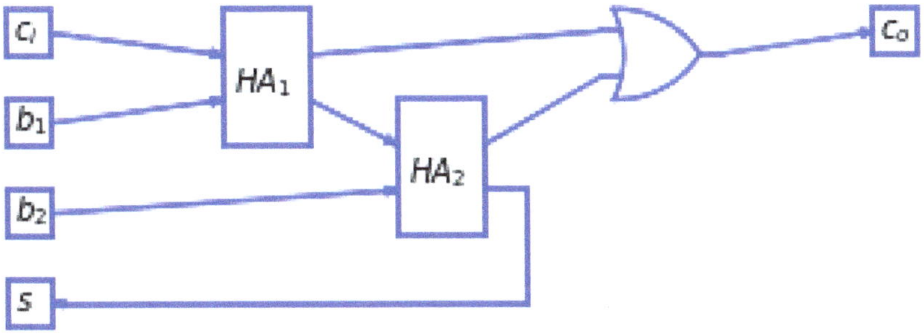

Figure 4.4: Full-Adder in Terms of Half-Adders

Inputs		Output
p	q	$p \wedge q$
0	0	0
0	1	0
1	0	0
1	1	1

Table 4.5: Logical AND

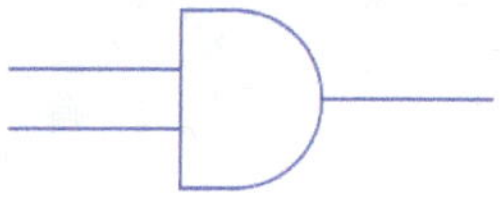

Figure 4.5: AND Gate

We will denote the binary logical operator AND by the symbol $\wedge$. Given the boolean values p and q, $p \wedge q$ will be 1 exactly when both p and q are equal to 1 and for all other combinations $p \wedge q$. This gives us the Table 4.5. The AND-gate is shown in Figure 4.5.

In contrast to all of the other operators we've encountered, the NOT operator is not a binary operator but a unitary operator. We denote it by $\neg$. Given a boolean value p, $\neg p$ has the opposite value, that is $\neg 1 = 0$ and $\neg 0 = 1$. We display this in Table 4.6. It is denoted in electronic diagrams by the symbol in Figure 4.6.

The NAND operator, as its name would suggest, is a composite of the AND and NOT operators. We denote this binary operator by $\uparrow$. If p and q are boolean values, then $p \uparrow q = \neg(p \wedge q)$. Note that $p \uparrow q$ is 1 except when p and q are both 1.

p	$\neg p$
0	1
1	0

Table 4.6: The NOT operator

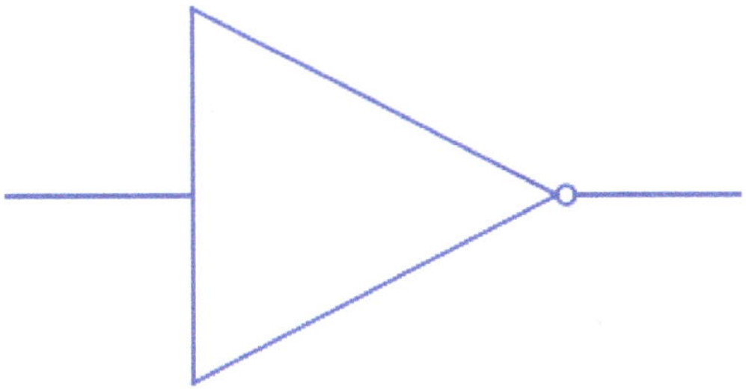

Figure 4.6: The NOT gate

p	q	$p \wedge q$	$\neg(p \wedge q)$	$p \uparrow q$
0	0	0	1	1
0	1	0	1	1
1	0	0	1	1
1	1	1	0	0

Table 4.7: Logical NAND

We build up in this operator in Table 4.7. The NAND gate is in Figure 4.7.

For some more insight into NAND, watch this video.

Our final composite operator is *exclusive OR*, i.e. XOR. We denote XOR by $\bigoplus$. Given boolean values p and q, $p \bigoplus q = 1$ exactly when one of p and q is equal to 1 and is 0 for all other combinations of p and q. We can write $p \bigoplus q$ in terms of the other operators as

$$p \bigoplus q = (p \vee q) \wedge \neg(p \wedge q).$$

This is justified in Table 4.8. The XOR gate is in Figure 4.8.

For those who've worked with modular arithmetic, it is quite natural to identify the boolean values 0 and 1 with the elements of $\mathbb{Z}_2 = \{0, 1\}$ of the fields of integers modulo 2. This is actually quite useful in the case of AND and XOR the operations denoted as $\wedge$ and $\bigoplus$ are equivalent to multiplication and addition in that setting.

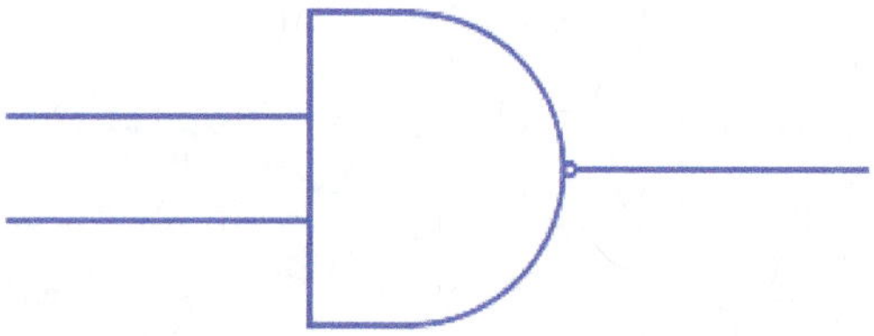

Figure 4.7: NAND Gate

p	q	$p \vee q$	$p \wedge q$	$(p \vee q) \wedge \neg(p \wedge q)$	$p \oplus q$
0	0	0	0	0	0
0	1	1	0	1	1
1	0	1	0	1	1
1	1	1	1	0	0

Table 4.8: Logical XOR

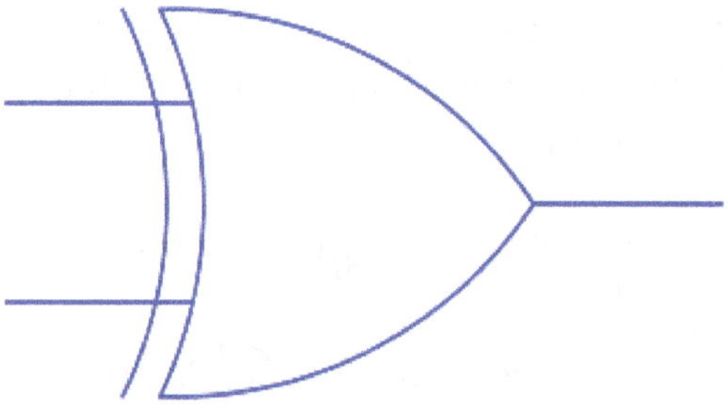

Figure 4.8: XOR-gate

b_1	b_2	$b_1 \oplus b_2$	$b_1 \wedge b_2$	s	c_{out}
0	0	0	0	0	0
0	1	1	0	1	0
1	0	1	0	1	0
1	1	0	1	0	1

Table 4.9: A Comparison of XOR, AND, and the outputs of the half-adder

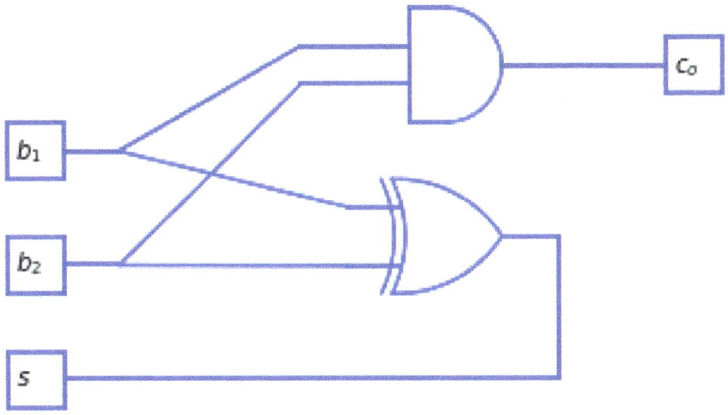

Figure 4.9: Half-Adder in terms of Logic Gates

4.8 The Half-Adder, Revisited

In this section, we will obtain a diagram for the half-adder in terms of basic logic gates. This will be made easy with the help of Table 4.9, which is quite evocative. Note that $b_1 \oplus b_2$ is exactly the same as s, the so-called sum of the half-adder. Also note that $b_1 \wedge b_2$ is exactly the same as c_{out}, the carry-out of the half-adder.

Using this information and the associated logic gates, we can wire up our half-adder as in Figure 4.9

Note that in diagrams such as these, we assume that crossing line segments do not interact unless there is a thick dot at the point of intersection. There are no such thick dots in this diagram.

p	q	$\neg p$	$\neg q$	$p \wedge q$	$\neg(p \wedge q)$	$\neg p \vee \neg q$
0	0	1	1	0	1	1
0	1	1	0	0	1	1
1	0	0	1	0	1	1
1	1	0	0	1	0	0

Table 4.10: Truth Table for Equation 4.5

This how leaves us with the task of writing the logic gates in terms of transistors. This is a two-fold process: First we will show that AND, OR, and XOR can be written in terms of NAND; then we will show that NAND can be written in terms of transistors. For the time being, we will hide the second of these behind a veil as would a skilled exotic dancer. The first requires a little more logic.

4.9 Just a Little More Logic

We will now present some formulas form logic that will be of great use to us. Their truth can be verified by creating truth tables for them. We will do this ourselves in one case and leave the rest to the interested reader.

Absorbtion:

$$p \vee p = p \tag{4.1}$$

$$p \wedge p = p \tag{4.2}$$

Involution:

$$\neg(\neg p) = p. \tag{4.3}$$

De Morgan's Laws:

$$\neg(p \vee q) = \neg p \wedge \neg q \tag{4.4}$$

$$\neg(p \wedge q) = \neg p \vee \neg q \tag{4.5}$$

To verify Equation 4.5, consider the truth table in Table 4.10.

We will now talk about the following identities where we put NOT, AND, OR, and XOR in terms of NAND:

1. $\neg p = p \uparrow p$

2. $p \wedge q = (p \uparrow q) \uparrow (p \uparrow q)$

3. $p \vee q = (p \uparrow p) \uparrow (q \uparrow q)$

4. $p \bigoplus q = (p \uparrow (p \uparrow q)) \uparrow (q \uparrow (p \uparrow q))$

We will demonstrate these identities. All of our demostrations are straightforward calculations with the exception of (4).

To show (1) we make use of absorbtion to show:

$$
\begin{aligned}
\neg p &= \neg(p \wedge p) \\
&= p \uparrow p
\end{aligned}
$$

We use involution and (1) to show:

$$
\begin{aligned}
p \wedge q &= \neg\neg(p \wedge q) \\
&= \neg(p \uparrow q) \\
&= (p \uparrow q) \uparrow (p \uparrow q)
\end{aligned}
$$

We can now apply absorbtion, (4.5), and (3) to show:

$$p \vee q = \neg(\neg p \wedge \neg q) \tag{4.6}$$

$$= \neg p \uparrow \neg q \tag{4.7}$$

$$= (p \uparrow p) \uparrow (q \uparrow q) \tag{4.8}$$

Making a truth table is the easiest way to verify (4), and we leave this to the reader. Using the forumlas (1) - (4), we can create diagrams for AND, OR, and XOR in terms of NAND in Figures 4.10-4.12.

Mutter to yourself "this is the easy way" while watching this.

4.10 Transistors and NAND

We now come to the last section which is the least mathematical part of the chapter. We will talk about transistors. To be precise, we are not talking about specific transistors, but about mathematically idealized transisitors. To talk about specific transistors would require discussion of specific voltages, specific resistances, etc. These details belong properly to the world of electronics or, dare we say, electrical engineering. We won't go to that world–at least in this chapter.

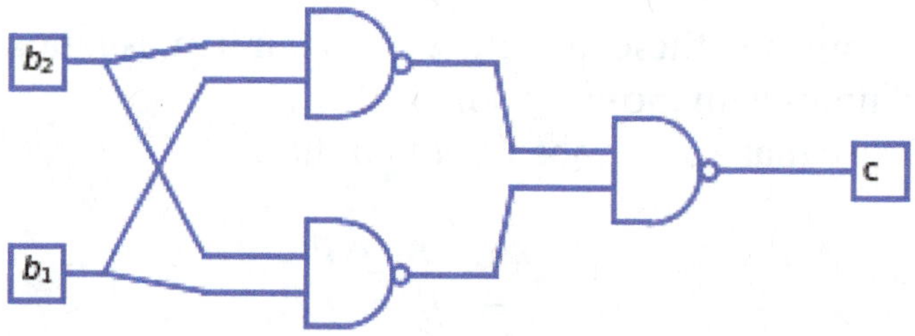

Figure 4.10: AND in terms of NAND with inputs b_1 and b_2

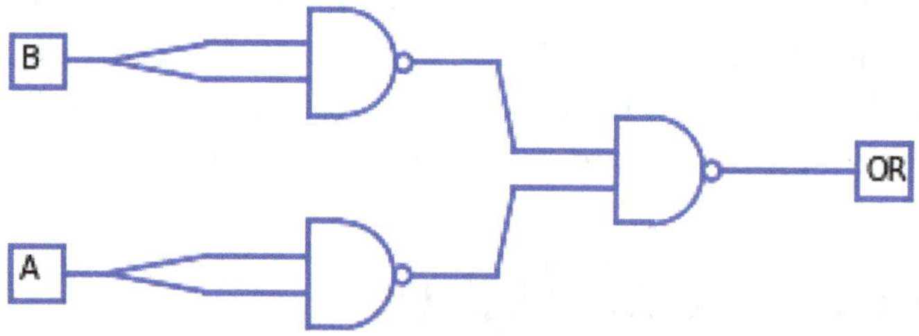

Figure 4.11: OR in terms of NAND with inputs A and B

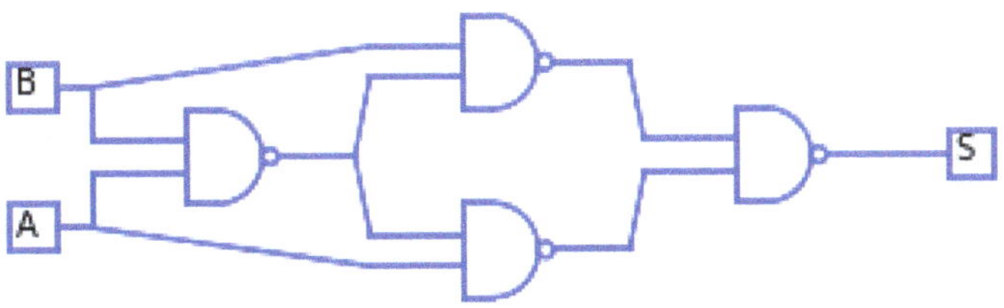

Figure 4.12: XOR in terms of NAND with inputs A and B

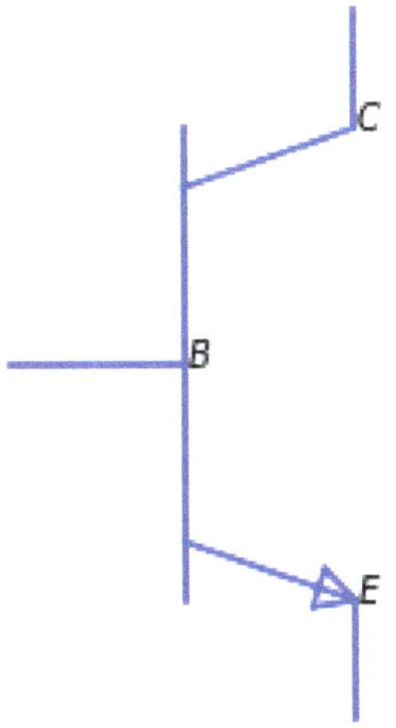

Figure 4.13: Mathematical npn Transistor

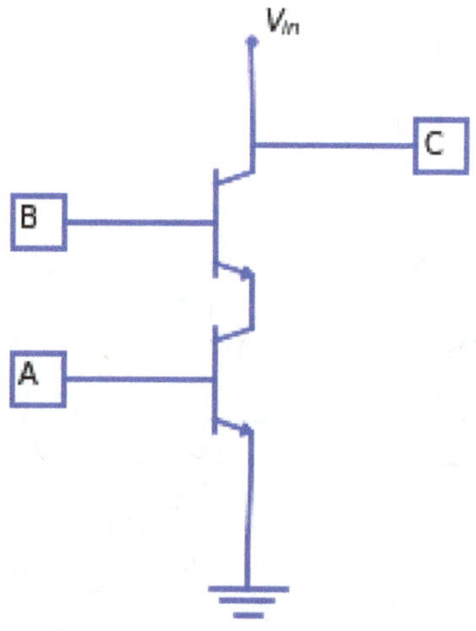

Figure 4.14: A NAND gate in terms of npn Transistors

Indeed, we are going to limit ourselves to exactly one type of transistor: a mathematically idealized npn transistor. See Figure 4.13. In the figure, there are three distringuished points: C, B, and E. We will call C the *collector*, E the *emitter*, and B the *base*.

For pedagogical purposes, the collector and the emitter are well-named. The collector is the place where electricity goes in, the place that collects it. The emitter is the place where electricity goes out, the place that emits it.

Pedagogically, it would have been better to refer to the base as the button. If you apply a particular voltage to the base, the transistor will then allow current to pass from the collector to the emittor. A voltage at B allows current to go from C and pass out of the transistor through E. A transistor is an electronic switch.

Now consider Figure 4.14. There is voltage applied to point V_{in}. In the diagram, proceeding downward from V_{in}, there is a branch point where our diagram separates into two paths: one horizonal; one vertical. At this point the current could go either horizontally to C or it could proceed down vertically through two transistors that are connected in series.

We will imagine the resistance from the branch point to C to be a bit greater

A	B	C	$A \uparrow B$
0	0	1	1
0	1	1	1
1	0	1	1
1	1	0	0

Table 4.11: C is $A \uparrow B$

than that of the series of transistors when both of them are activated. [3]

If A and B are both 1, then the current from V_{in} will be allowed to go to ground. This means there will be insufficient voltage at C, i.e. $C = 0$. Otherwise, $C = 1$ because if either A or B is equal to 0 current will not be allows to pass through the series of transistors. This is written in detail in Table 4.11. We see that the value at C is $A \uparrow B$.

4.11 Epilog

We are at the end of our journey. In principle, you could build an electronic device to add numbers. This would require using transistors to make NAND gates: NAND gates to make the various other logic gates; and the various other logic gates to make half-adders and full-adders. When it is all put together, it is a very complicated device.

There have been sturdy individuals who've done just that. I doff my hat to them. For my part, I am in a state of awe to those who created the means of tackling such complexity.

[3] I *resisted* the temptation to talk about resistors.

4.12 Exercises for Chapter 4

1. Write the following binary numbers in their expanded form, *i.e.* as $b = 2^n \cdot b_n + \cdots + 2 \cdot b_1 + b_0$.

   (a) $(1011)_2$

   (b) $(10111)_2$

   (c) $(1111101)_2$

2. How many bits are in each of the following numbers.

   (a) $(1011)_2$

   (b) $(10111)_2$

   (c) $(1111101)_2$

3. What is the maximum value of a binary number will the given number of bits? Express your answer in decimal.

   (a) $N = 5$

   (b) $N = 10$

   (c) $N = 16$

   (d) $N = 32$

4. Perform the additions of binary numbers by hand. Leave the answer in binary.

   (a) $(110)_2 + (1011)_2$

   (b) $(1001)_2 + (1111)_2$

   (c) $(100111)_2 + (111001)_2$

   (d) $(101111)_2 + (111101)_2$

5. In the following, use a half-adder to perform the addition of bits to obtain s and c putting all inputs and outputs in your version of Figure 4.1.

   (a) Let $b_1 = 0$ and $b_2 = 1$.

   (b) Let $b_1 = 1$ and $b_2 = 0$.

   (c) Let $b_1 = 1$ and $b_2 = 1$.

6. In the following, use a full-adder to perform the addition of bits to obtain s and c_{out} putting all inputs and outputs in your version of Figure 4.1.

   (a) Let $c_{\text{in}} = 1$, $b_1 = 0$ and $b_2 = 1$.

   (b) Let $c_{\text{in}} = 1$, $b_1 = 1$ and $b_2 = 0$.

   (c) Let $c_{\text{in}} = 1$, $b_1 = 1$ and $b_2 = 1$.

7. Express $p \wedge (q \vee r)$ in a diagram using logic gates.

8. Express $(p \wedge q) \vee (p \wedge r))$ in a diagram using logic gates.

9. Express $\neg(p \wedge q)$ in a diagram using logic gates.

10. Express $\neg p \vee \neg q$ using logic gates.

11. Make truth tables for the following.

    (a) $p \bigoplus q$

    (b) $\neg p \bigoplus \neg q$

    (c) $\neg(p \bigoplus q)$

    (d) How does $\neg p \bigoplus \neg q$ relate to $\neg(p \bigoplus q)$.

12. Make truth tables for the following.

    (a) $(p \uparrow q) \uparrow p$

    (b) $\neg p \vee q$

13. Create truth tables to verify each of the following.

    (a) $p \vee p = p$

    (b) $p \wedge p = p$

    (c) $\neg(\neg p) = p$

14. Demonstrate $p \bigoplus q = (p \uparrow (p \uparrow q)) \uparrow (q \uparrow (p \uparrow q))$ by means of a truth table.

Chapter 5

More Numbers

5.1 Bit-wise Operations

While numbers in mathematics can be of any size, when using computers, we are constrained to be responsible. In computers, as numbers are represented in bits, a fixed number of bits is chosen to hold a particular number or data type. This number of bits of bits varies according to the machine and type of number, but typically, it is some power of two such as 4, 8, 16, 32, 64, or 128. In this section, we will denote this number of bits by N. We will be considering operations that be done on the set of binary numbers of length N[1].

In Chapter 4, we became acquainted with a variety of operations, both unitary and binary, on $\mathbf{Z}_2$ the set of boolean numbers: $\neg$, $\vee$, $\wedge$, $\oplus$, and $\uparrow$. We will presently examine how these can be extended to the set of binary numbers of length N.

We will begin this program with the $\neg$ operator. For notational reasons that will soon become clear, let us denote $\neg x$ by $\bar{x}$. Given a binary number of length N,

$$a = (a_{N-1}a_{N-2}\dots a_1 a_0)_2,$$

define

$$\begin{aligned} \neg a &= \bar{a} \\ &= (\bar{a}_{N-1}\bar{a}_{N-2}\dots \bar{a}_1 \bar{a}_0)_2. \end{aligned}$$

[1]The mathematician in me is yearning to denote this by the symbol $\mathbf{Z^N}$ and start talking about vector spaces and the like, but you will have time enough to learn about that on your own.

So, for example, with $N = 4$,

$$\begin{aligned} \neg(10)_{10} &= \neg(1010)_2 \\ &= (\bar{1}\bar{0}\bar{1}\bar{0}) \\ &= (0101)_2 \\ &= (5)_{10}. \end{aligned}$$

The careful reader might note that using a different length N will yield different results. For $N = 8$,

$$\begin{aligned} \neg(10)_{10} &= \neg(00001010)_2 \\ &= (11110101)_2 \\ &= (245)_{10}. \end{aligned}$$

There are some special cases where this yields interesting results. For example, for a given N,

$$\begin{aligned} \neg 0 &= \neg\underbrace{(00\ldots00)_2}_{N \text{ zeros}} \\ &= \underbrace{(\bar{0}\bar{0}\ldots\bar{0}\bar{0})_2}_{N\ \bar{0}\text{s}} \\ &= \underbrace{(11\ldots11)_2}_{N\ 1\text{s}} \\ &= 2^N - 1 \end{aligned}$$

because

$$\begin{aligned} 2^N - 1 &= (2-1)(2^{N-1} + 2^{N-2} + \cdots + 2 + 1) \\ &= (11\ldots11)_2. \end{aligned}$$

We may, in a similar way, extend any binary operation on $\mathbf{Z}_2$ to one on binary numbers of length N. For example consider the OR operation $\vee$. Given binary numbers of length N, $a = (a_{N-1}a_{N-2}\ldots a_1a_0)_2$ and $b = (b_{N-1}b_{N-2}\ldots b_1b_0)_2$, define

$$a \vee b = (c_{N-1}c_{N-2}\ldots c_1c_0)_2,$$

where $c_i = a_i \vee b_i$ for each i.

To illustrate this in the case $N = 4$, consider

$$\begin{aligned} 5 \vee 9 &= (0101)_2 \vee (1001)_2 \\ &= (1101)_2 \\ &= 13. \end{aligned}$$

One might note that 0 acts as an identity element for $\vee$ as

$$\begin{aligned} 0 \vee a &= (00\ldots00) \vee (a_{N-1}a_{N-2}\ldots a_1 a_0)_2 \\ &= (a_{N-1}a_{N-2}\ldots a_1 a_0)_2 \\ &= a \end{aligned}$$

for any a.

We may also extend $\wedge$ to operate on binary numbers of length N in the very same way:

$$\begin{aligned} a \wedge b &= (a_{N-1}a_{N-2}\ldots a_1 a_0)_2 \wedge (b_{N-1}b_{N-2}\ldots b_1 b_0)_2 \\ &= (c_{N-1}c_{N-2}\ldots c_1 c_0)_2 \end{aligned}$$

where $c_i = a_i \wedge b_i$ for all i.

To illustrate this in the case $N = 4$, note

$$\begin{aligned} 5 \wedge 9 &= (0101)_2 \wedge (1001)_2 \\ &= (0001)_2 \\ &= 1. \end{aligned}$$

There is a special case of this that is interesting to us. Consider

$$\begin{aligned} 1 \vee a &= \underbrace{(00\ldots00)_2}_{N \text{ zeros}} \wedge (a_{N-1}a_{N-2}\ldots a_1 a_0)_2 \\ &= (\underbrace{0\ldots0}_{N-1 \text{ zeros}}\ a_0)_2 \\ &= a_0 \\ &= (a \,\%\, 2). \end{aligned}$$

That is to say, "anding" a with 1 picks off the least significant bit of a.

Figure 5.1: Shift Register

5.2 Shifts

In addition to the bit-wise operations we looked at in Section 5.1, there are other operations that can be performed on binary numbers of a fixed length N. One class of such operations are *shifts.* We will examine those in this section, both *left shifts* and *right shifts.*

For the purposes of this section, it is useful to think of an N-bit binary number being stored in a series of adjacent mailboxes as in Figure 5.1. Let us now consider a new, and not necessarily too bright, employee from the mail room as he comes to deliver a 0 to the rightmost mailbox. He goes to put a 0 there, but finding it already filled with a_0, he shifts a_0 to the next box immediately to the left. Finding that filled with a_1, he shifts it one box to the right, and so on until he comes to the last box which is filled with a_{N-1}. He puts a_{N-2} into that box, and throws a_{N-1} away. This leaves the mail boxes in the state we see in Figure 5.2. This operation is called a *left shift,* for obvious reasons. We will denote by λ and define it formally by

$$\lambda((a_{N-1}a_{N-2}\ldots a_1a_0)_2) = (a_{N-2}a_{N-3}\ldots a_00).$$

In the case that $a_{N-1} = 0$, this shift accomplishes the mathematical operation

Figure 5.2: Left Shift

of doubling, *i.e.* multiplication by 2. We can see this below as

$$\begin{aligned}\lambda(a) &= \lambda((a_{N-1}a_{N-2}\ldots a_1 a_0)_2)\\ &= \lambda((0a_{N-2}\ldots a_0)_2)\\ &= (a_{N-2}\ldots a_0 0)_2\\ &= a_{N-2}2^{N-1}+\cdots+a_0 2\\ &= 2(a_{N-2}2^{N-2}+\cdots+a_0)\\ &= 2a.\end{aligned}$$

In order to define right shifts, let us return our attention to our hapless mail room employee. Having been criticized for delivering mail as he'd done before, he now delivers his 0 to the left most box and shifts the contents of each of the other boxes one box to the right until he has to throw a_0 away. This will yield mail boxes that look like the picture in Figure 5.3.

This, understandably, is referred to as a *right shift* and is denoted by rho so that

$$\begin{aligned}\rho((a_{N-1}a_{N-2}\ldots a_1 a_0)_2) &= = (0a_{N-1}\ldots a_1)_2\\ &= a_{N-1}2^{N-2}+\cdots+a_1\\ &= (a//2).\end{aligned}$$

Figure 5.3: Right Shift

It is also useful to observe that

$$\rho^k((a_{N-1}a_{N-2}\ldots a_1 a_0)_2) \wedge 1 = a_k.$$

The effect of this is that we can use a combination of ρ and $\wedge$ to pick out any particular bit of a.

5.3 Multiplication

5.3.1 Introduction

In Chapter 4, we spent a great deal of time learning how to add, as it were. This required a great deal of work and the (mathematical) wiring together of many transistors. Having already invested that much effort, it would be good if we could put it to greater use.

In this section, it is our aim to do just that. We will find ways to us addition to accomplish both multiplication and subtraction. While the methods we use are not needlessly complicated, they may not be the ones of which the reader is currently thinking. We invite the reader to come along as we examine them, and shall tackle multiplication first.

```
Input m, n
Set c = 0
Set A = 0
While c < m
        A = A + n
        c = c + 1
return A = nm
```

Figure 5.4: Naive Multiplication Algorithm

5.3.2 The Naive Multiplication Algorithm

From the very beginning of our mathematical lives, we are taught that multiplication (of integers at least) is repeated addition. Because of this, there is a naive algorithm the reader might be able to put together without any help. We include it here for the purpose of reference and comparison.

Let m and n be positive integers. Choose notation so that $m \leq n$. Let c be initialize to the value 0 so that we can use it to count how many times we go through a loop. Let A be initialized as 0 to be a place in which we can accumulate our results.

Perform the following operations: Add n to A and store the results in A; increment c by 1. Repeat these steps as along as $c < m$. When $c = m$, it will be the case that $A = nm$. This can be implemented in pseudo code as in Figure 5.4. This algorithm requires m additions. While this is not all that bad, we will, in fact, be able to do better. In Section 5.3.3, we will introduce an algorithm that is considerably less naive that this one and quite a bit faster: We will be able to multiply m and n with approximately $\log_2 m$ additions.

5.3.3 The Russian Peasant Algorithm

As the following algorithm has a long history in recreational mathematics, we will respect that by introducing it in a playful way. Given two integers m and n, we will make two piles of beans: One pile will contain m beans and the other will contain n beans. Again we may assume $m \leq n$, but this need not be the case for this method to work properly.

Let us imagine out two piles of beans in the following way. Imagine each pile is contained in a bowl and the two bowls are sitting on a long, narrow table.

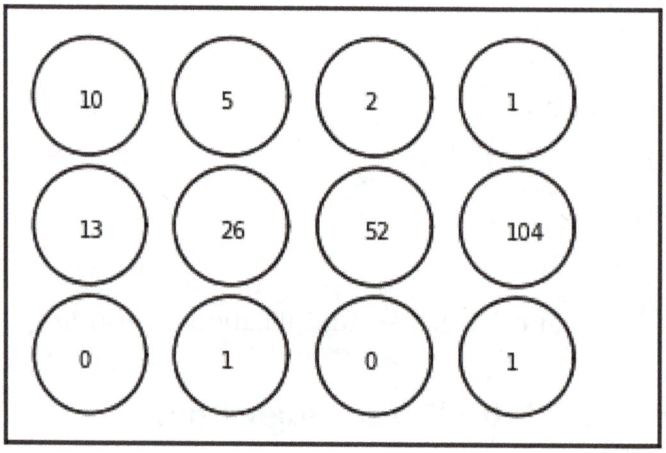

Figure 5.5: Bowls of Beans on the Table

They are in a row which is parallel to the narrow side of the table. Along side them, there is a third bowl in the same row. It is either empty or contains a single bean, depending upon the number of beans in the first bowl. If the first bowl contains an even number of beans, it will be empty; if the first bowl contains an odd number, it will contain a single bean.

Let's illustrate this with specific values. Say $m = 10$ and $n = 13$. Then the first bowl contains 10 beans, the second 13, and the third none. For the next round we need three more bowls that will be set up parallel to the first three.

In this round, we determine the number of beans in the new first bowl by looking to the previous round. We cut that number in half and put 5 beans into that bowl. In the second bowl, we look at the corresponding bowl from the previous round and put in twice as many beans, *i.e* 26. In the third bowl we put in a single bean because 5 is odd.

We continue on in this way laying out our rows of bowls: halving the first; doubling the second; putting either no bean or a single bean in the third depending upon whether the number of beans in the first plate is even or odd, respectively.

The process stops when we have a single bean in the first bowl.

We can see a representation of our beans on the table in Figure 5.5 We summarize this in Table 5.1

1st Bowl	2nd Bowl	3rd Bowl
10	13	0
5	26	1
2	52	0
1	104	1

Table 5.1: Summary of Bean Bowls

Halve	Double	Remainder
$a_n 2^n + \cdots + a_1 2 + a_0$	b	a_0
$a_n 2^{n-1} + \cdots + a_1$	$2b$	a_1
$\vdots$	$\vdots$	$\vdots$
$a_n 2 + a_{n-1}$	$2^{n-1}b$	a_{n-1}
a_n	$2^n b$	a_n

Table 5.2: Mathematics, not Beans

We now focus our attention on the second bowl from each stage. Indeed, we only want the second bowls where the third bowl in that stage contained a single bean. The beans from these bowls we all dump into an empty bowl and then count the total. From this example it will be 26 + 104 which total to 130. And it happens that $130 = 10 \cdot 13 = mn$. This is not an accident.

Let us now rerun this experiment with more mathematics and fewer beans.

Let $a = a_n 2^n + \cdots + a_1 2 + a_0$ be the first number and let b be the second. We use these numbers to fill out the columns of a table without counting the beans. See Table 5.2 Note that the remainders are just the least significant bit in the binary expansion of the numbers in the first column. Let us now consider the sum of the products of the "Double" column and the "Remainder" column, more specifically:

$$ba_0 + 2ba_1 + \cdots + 2^n ba_n. \tag{5.1}$$

The a_is are either 0 or 1. If $a_i = 0$, then of course $2b^i a_i = 0$. So the expression in (5.1) is equal to the sum of the entries in the "Double" column for which there is a 1 in the "Remainder" column, the sum we have been computing in

Input m, n
$c = 0$
While$\neg(m == 0)$
    If $m \wedge 1 == 1$
        $c = c + n$
    $n = \lambda(n)$
    $m = \rho(m)$
Return c

Table 5.3: Pseudo-code for Russian Peasant Algorithm

our game. Note

$$\begin{aligned} ba_0 + 2ba_1 + \cdots + 2^n ba_n &= b(a_0 + \cdots + 2^n a_n) \\ &= ab \end{aligned}$$

The process we've described in this section is called "The Russian Peasant Algorithm." It provides us a way to multiply numbers that allows us to make use of the binary format in which a computer stores those numbers. This will be made more succinct in the next section.

5.3.4 Grown-up Russian Peasant Algorithm

Let us now approach the Russian Peasant Algorithm in a less colorful way. We describe the algorithm in pseudo-code in Table 5.3. Observe that $m \wedge 1$ is the rightmost bit from m (see Section 5.1); that λ is the left shift and so multiplies by 2; and that ρ is the right shift and so it integer division by 2. This algorithm will terminate when m is equal to 0.

The algorithm implemented in this way requires a number of shifts equal to the number of bits in m. This is approximately $\log_2 m$. As a consequence, there will be at most $\log_2 m$ additions needed to multiply. Since $\log_2 m < m$, this algorithm is an improvement over the Naive method and, in most cases, a considerable improvement.

5.4 Subtraction

5.4.1 Introduction

We now turn or attention to learning how to subtract. Again we wish to leverage the work we've done already in creating addition. Because of this, we will be using addition in order to subtract. This is not really a new idea as from our school math course we know $a - b = a + (-b)$. In essence we will be creating ways to find $-b$ or something that can be made to act like $-b$. We will actually find two such things: The ones complement and the twos complement.

5.4.2 Ones Complement

In each of our subtraction methods, we will first perform a special case of subtraction from a particular fixed number. As in earlier sections, we will be working in a context of each of our numbers having a fixed number N of bits.

Fix $N > 1$ to be an integer. Let $B = 2^N - 1$. This is the largest number that can be represented by N bit. It is written as

$$B = \underbrace{(11\ldots11)_2}_{N\ 1\text{s}}.$$

Let $(b_{N-1}b_{N-2}\ldots b_1 b_0)_2$ be an N-bit binary number. As b is necessarily no bigger than b on a bit by bit basis, subtracting b from B is quite easy. Let us look at an example when $N = 4$. Let $b = (0101)_2$. Then

$$\begin{aligned} B - b &= 1111 - 0101 \\ &= 1010. \end{aligned}$$

Notice that the overall effect is to change the bits in b that are 0 to 1 and those that are 1 to 0. This is exactly what happens when we apply $\neg$ to b bit wise as we did in Section 5.1. Using the notation we learned in that section, we see

$$B - b = \bar{b}.$$

In this context $\bar{b}$ is called the ones complement of b. We display $\bar{b}$ for values of b from 0 through 7 in Table 5.4. Using 4-bits, we can represent the integers 0 through 15, however, when the leftmost bit is equal to 1 in this context, it is representing a negative number. The numbers whose leftmost bit is 1 represent

b	4-bit binary	$\bar{b}$	$-b$
0	0000	1111	-0
1	0001	1110	-1
2	0010	1101	-2
3	0011	1100	-3
4	0100	1100	-4
5	0101	1010	-5
6	0110	1001	-6
1	0111	1000	-7

Table 5.4: Ones Complement

-7 through 0. The particularly attentive reader will have noticed we have two very different ways to represent 0: 0000 and 1111. This consideration is a part of the price of this method. Let us do some analysis to see how this works.

Let a and b be positive integers than can be represented in $N-1$ bits. Let us consider how to calculate $a-b$. There are three cases to consider: $a=b$, $a<b$, and $a>b$. In each of these cases, we will be (kind of) using $\bar{b}$, the ones complement of b as $-b$ and considering $a+\bar{b}$.

Let us first consider the case $a=b$. Then

$$a+\bar{b}=a+\bar{a}.$$

When the ith bit of a is 1 the ith bit of $\bar{a}$ will be 0 and vice versa. Consequently,

$$\begin{aligned} a+\bar{a} &= \underbrace{1\ldots11}_{N\text{ 1s}} \\ &= B. \end{aligned}$$

Recall that B is negative zero, which is to say zero. So our system is consistent.

Now consider the case $a<b$. Then

$$\begin{aligned} a+\bar{b} &= a+B-b \\ &= B-(b-a). \end{aligned}$$

Since $b>a$, it follows that $b-a$ can be represented using only the rightmost $N-1$ bits of the system. (Recall our assumption that both a and b can be represented by $N-1$ bits.) In this case, $B-(b-a)$ is equal to $b\bar{-}a$, the ones complement of $b-a$. Therefore,

$$a+\bar{b}=b\bar{-}a.$$

Since $a < b$, $a - b = -(b - a)$. So once again our system is consistent.

Let's look at a concrete example to see how this is playing out. Let $a = 3$ and $b = 7$. Then in 4-bit binary, $a = 0011$ and $\bar{b} = 1000$. It follows that $a + \bar{b} = 1011$. As the leftmost bit is 1, we interpret this to be a negative number. It is the ones complement of 0100 which is the 4-bit binary form of 4. Therefore, we interpret 1011 as -4, which is the correct answer.

This worked in a straightforward manner. Let us now consider the case $a > b$. Then

$$\begin{aligned} a + \bar{b} &= a + B - b \\ &= B + (a - b) \\ &= B + 1 + (a - b - 1) \\ &= 2^N + (a - b - 1). \end{aligned}$$

Since $a > b$ and a and b are both integers, $a - b - 1 \geq 0$. This implies the binary form of $2^N + (a - b - 1)$ will require $N + 1$ binary bits to be represented. This is an overflow of our N-bit addition system. One practical effect of this is that the value of the carry-out of the leftmost full-adder of our addition processor (our adder) will be equal to 1. While we could just dismiss this as a disaster and give up, let us not lose hope just yet. Indeed, let us revisit our calculation that

$$a + \bar{b} = 2^N + (a - b - 1).$$

If we use modular arithmetic from Section 1.5.4, we see

$$a + \bar{b} \equiv_{2^N} a - b - 1.$$

To the fainthearted and unimaginative, this is just a wrong answer, because the answer should be $a - b$ not $a - b - 1$, but to the bold, it is workable. Why not simply add 1 to the result? In fact, that is what we do.

Let us revisit our work on addition from Chapter 4 and use the black box representation in Figure 5.6 to represent the addition of N-bit numbers a and b. Here s is the "attempted" sum of the two numbers. I use the quote marks because it is possible the sum of the two numbers is too large to be contained in N bits, in which case there is an overflow which will result in the last carry-out bit being equal to 1. In our figure, we label this last carry-out by the letter o so as to stand for *overflow*. We will either have $o = 1$ or $o = 0$.[2]

We can construct a subtracter by joining two adders together with a well-placed inverter. We've done such in Figure 5.7. When $a > b$, s will have 0 as its

[2]The choice of font is very important here. Should we use ω instead?

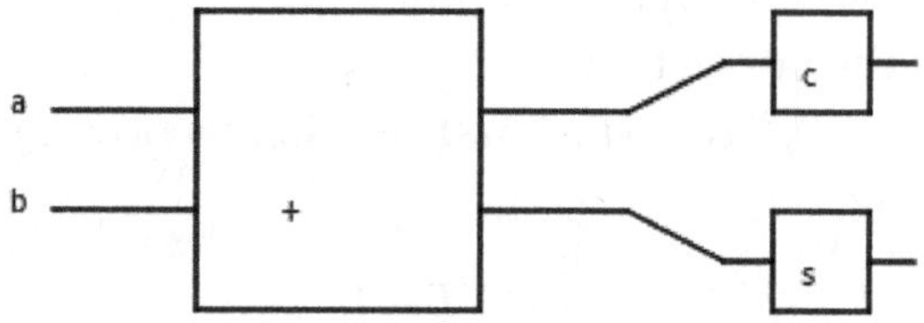

Figure 5.6: Adder

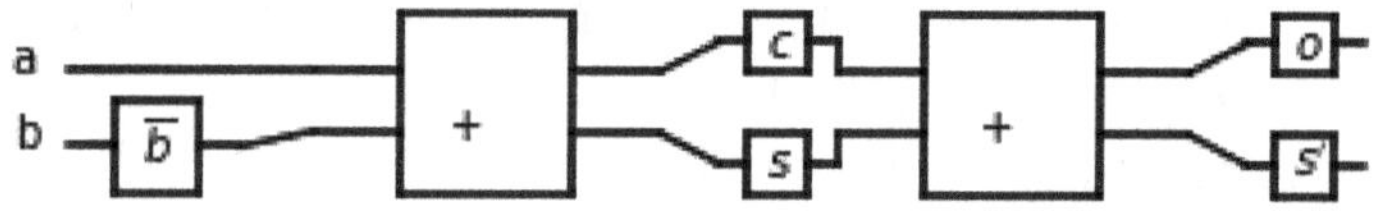

Figure 5.7: Subtracter

leftmost bit and be the binary representation of $a-b$. When $a<b$, s' will have 1 as its leftmost bit and be equal to the ones complement of $b-a$.

For example, let $a=7$ and $b=3$. Then $a=(0111)_2$ and $\overline{b}=(1100)_2$. So $a+\overline{b}=(10011)_2$, which is an overflow of our 4-bit system (*i.e.* $c=1$). When we add 1 to s, the truncation of (10011)2 to $(0011)_2$ this will give us $(0100)_2$. As the left most bit of $(0100)_2$ is 0, we interpret it to be a positive number so that $7-3=4$.

5.4.3 Twos Complement

We will now introduce another way to leverage addition to perform subtraction: *the twos complement.* While it is slightly more complicated to determine the twos complement than it is the ones complement, this method avoids having non-unique representations of zero.

Recall we began the ones complement method by subtracting a number b from $B=2^N-1$ in our N-bit system. In the twos complement method, by way of contrast, we will be subtracting b from 2^N itself. That is to say, the twos complement of b is

$$2^N-b.$$

We will denote the twos complement of b as $\widetilde{b}$, *i.e.* $\widetilde{b}=2^N-b$. Note that

$$\begin{aligned}\widetilde{b} &= 2^N-b\\ &= (2^N-1)-b+1\\ &= B-b+1\\ &= \overline{b}+1.\end{aligned}$$

That is to say, the twos complement can be obtained from the ones complement by adding 1 and ignoring any carry to the $(N+1)$st bit. See Table 5.5. We can also obtain the twos complement of b by holding the rightmost 1 and all other bits to its right fixed and applying the NOT operator to all of the bits to the left.

As in Section 5.4.2, we can perform the subtraction of b from a by adding a to the twos complement of b:

$$\begin{aligned}a+\widetilde{b} &= a+(2^N-b) &\quad (5.2)\\ &= 2^N+a-b &\quad (5.3)\\ &\equiv_{2^N} a-b. &\quad (5.4)\end{aligned}$$

b	binary	ones complement	twos complement
0	0000	1111	0000
1	0001	1110	1111
2	0010	1101	1110
3	0011	1100	1101
4	0100	1011	1100
5	0101	1010	1011
6	0110	1001	1010
7	0111	1000	1001

Table 5.5: Twos Complement

The equivalence modulo 2^N takes into account ignoring any carry into the $(N+1)$st bit.

Let us now work a couple of concrete examples so that we can see the wheels turning on this process.

Consider $3-7$. To begin, we represent 3 in 4-bit binary as $(0011)_2$ and 7 as $(0111)_2$. The twos complement of 7 is (1001). When we add $(0011)_2$ to $(1001)_2$, we get $(1100)_2$. As the leftmost bit is 1, we know that this represents a negative number in this notation, so we takes its twos complement: $(0100)_2$. This is the binary representation of 4. So $3-7=-4$. Mystery solved.

Let us now consider $7-3$. Using the binary forms from the previous paragraph, we find the twos complement of 3 is $(1101)_2$. By adding $(0111)_2$ to $(1101)_2$, we get $(10100)_2$ which is an overflow of our 4-bit system, but we ignore the overflow bit, giving us the answer $(0100)_2$ which is the binary representation of 4. Therefore, we've calculated $7-3=4$ and the world is safe.

To be complete, let us consider $(-2)+(-5)$. The twos complement of 2 is $(1110)_2$ and the twos complement of 5 is $(1011)_2$. These sum to $(11001)_2$, but in our 4-system this truncates to $(1001)_2$. As the leftmost bit in $(1001)_2$ is 1, we know this represents a negative number. Its twos complement is $(0111)_2$ which is the binary representation of 7. Therefore, $(1001)_2$ represents -7, the correct answer.

In Figure 5.8 we describe an algorithm to create the twos complement of a number directly from its bits.

Input $b_{N-1}, b_{N-2}, \ldots, b_1, b_0$
index = 0
while b[index] = 0
    index = index + 1
index = index + 1
while index $\leq N-1$
    b[index] = $\neg b$[index]
    index = index + 1
return $b_{N-1}, b_{N-2}, \ldots, b_1, b_0$

Figure 5.8: Algorithm to Compute the Twos Complement

www.ingramcontent.com/pod-product-compliance
Lightning Source LLC
LaVergne TN
LVHW060452061126
851857LV00016B/1801
9798385131655